DUTY
FIRST

Also by Ed Ruggero

DUTY
FIRST

*West Point and
the Making of
American Leaders*

ED RUGGERO

HarperCollins*Publishers*

Inquiries should be addressed to Permissions Department, HarperCollins Publishers Inc., 10 East 53rd Street, New York, NY 10022.

It is the policy of HarperCollins Inc., and its imprints and affiliates, recognizing the importance of preserving what has been written, to print the books we publish on acid-free paper, and we exert our best efforts to that end.

Library of Congress Cataloging-in-Publication Data

Ruggero, Ed.
    Duty first : West Point and the making of American leaders / by Ed Ruggero.—1st ed.
        p. cm.
    ISBN 0-06-019317-4 (alk. paper)
    1. United States Military Academy. 2. Leadership. I. Title.

U410.L1 R84 2001
355'.0071'173—dc21

                                                                    00-059775

Printed in the United States of America

FIRST EDITION

00 01 02 03 04 WB/RRD 10 9 8 7 6 5 4 3 2 1

www.harpercollins.com

to Marcia
for everything

CONTENTS

Duty, Honor, Country
West Point's motto

THE MISSION OF THE UNITED STATES MILITARY ACADEMY

To educate, train, and inspire the Corps of Cadets so that each graduate is a commissioned leader of character committed to the values of Duty, Honor, Country; professional growth throughout a career as an officer in the United States Army; and a lifetime of selfless service to the Nation.

Duty is the sublimest word in our language. Do your duty in all things. You cannot do more; you should never wish to do less.

Robert E. Lee
Class of 1829

INTRODUCTION

Lieutenant General Dan Christman, West Point's Superintendent, is a big, florid-faced man with a wide smile, an overpowering charm, and a very specific vision for the United States Military Academy. West Point, he will tell anyone who will listen, is "America's premier leadership school."

In 1998 I heard Christman use this phrase repeatedly during a three-day meeting with the presidents of West Point's regional alumni organizations. In that room, at least, Christman was preaching to the choir, and the choir already believed. Graduates know the names of the West Pointers who have shaped American history: Eisenhower and MacArthur, Grant and Lee, Pershing and Schwarzkopf and Patton. They also know of the scores of leaders who serve the nation in the military and, after their service, in a wide array of civilian professions.

At the twentieth reunion of the Class of 1980, for instance, a visitor could meet: a member of Congress, four people who have worked at the White House, the military attachés to Vietnam and Jordan, a shuttle astronaut (and space walk veteran), a heart surgeon, an eye surgeon, an FBI special agent, CEOs, physicians, university professors, ministers, lawyers, entrepreneurs, engineers, scientists, airline pilots. This would also be the place to get firsthand accounts of what it's like to command five hundred peacekeepers in the Balkans, or a battalion of Green Berets, or half a hundred attack helicopters on the DMZ between North and South Korea.

West Point may or may not be, in Christman's words, *the* school for leaders, but it is arguably among the best. If the successes of its graduates are any indicator, the Academy's approach offers a template for leader development in and out of the military. There certainly is a need for leadership. Most American institutions are, in the words of Harvard's John P. Kotter, overmanaged and under-led. Businesses spend millions on consultants who wheel in checklists and decision-matrices. Then the consultants go home and we find, to our constant surprise, that employees are still not inspired.

For two hundred years West Point has taken talented young Americans and put them through an intense four-year program to build leaders of character. On graduation day the Superintendent sends them out with a rolled-up diploma and an astonishing set of experiences. How do those experiences help mold leaders? I'd spent four years as a cadet and another four on the faculty (out of eleven years' service), and still wasn't sure I could put my finger on exactly how it happened. So in 1998 I started looking for an answer, and I began my search in the office of a leadership professor at USMA.

"If you ask five people around here how leaders are made, you're going to get five different answers."

Lieutenant Colonel Scott Snook delivers this not-quite-what-I-was-looking-for answer in his windowless office deep inside Thayer Hall, the Academy's largest academic building. Unlike the stereotypical professor's office, this one is neat: The books are arranged by subject, there are no piles of student papers or coffee-stained journals. One large bulletin board shows a military map of Grenada and a photograph of some GIs—armed, their faces dark with camou-flage—holding a Cuban flag. The soldier kneeling at the lower right is then-Lieutenant Snook. In the corner of another bulletin board is a movie still of John Wayne from the 1962 D-Day epic, *The Longest Day.* Snook's Harvard degrees—an M.B.A. and a Ph.D.—hang on another wall.

I met Scott Snook when he was a much-less-accomplished year-ling, or second-year cadet, and we were assigned to dig a foxhole together.

Snook rolls his chair to a filing cabinet and pulls a folder, then slides the packet across the desk insistently. The document inside, dark with close-spaced type, is the result of an in-depth study of leader development at West Point. The language, dense and pedan-tic, goes on for several mind-numbing pages. Then, a sentence in bold type: **"USMA has no clearly articulated 'learning model' or theory for how to develop leaders of character."**

I thought this a pretty serious omission for an institution charged with doing exactly that—at great expense to the taxpayer.

"We do what we do now because it has worked in the past," Snook says. But there is no master plan, no theory to help determine what does and does not contribute. This explains why old grads (anyone from the most recently graduated back to eighty-year-old alumni) can talk about the same rite of passage, and one will claim, "It made me the man I am today"; while the other will say, "It was mostly stupid, fraternity-row stuff and a waste of my time."

"Think of it like an academic course," Snook says. "The Cadet Leader Development System is the syllabus. It describes what you do throughout the forty lessons of the semester. Then you have the tests and exams and papers to evaluate the student's understanding. We have all that in place, too. What we don't have is what comes before the syllabus, a theory of how students learn the subject."

This finding was not well received by the Commandant, the one-star general responsible for cadets' military training. Was it possible, the Commandant wanted to know, that this self-described "premier leadership institution" had merely stumbled onto something that had worked well for so long? How, exactly, does West Point develop leaders of character?

Snook's group wrote, "Our typical response is descriptive at best: 'We have three programs: the military, the academic, and the physical. Within each program, cadets participate in a series of progressive and sequential activities. Here is a list of those activities . . .' "

The study doesn't claim that what West Point is doing is flawed, but without a clearly articulated theory of how leaders are developed, there is no yardstick for evaluating new programs, no measure by which to judge current practices. The lack of an underlying theory means that questions about how to do things—and what the right things are—are difficult to address.

The report goes on to say that, "If we believe that the West Point experience is fundamentally sound, then we should be able to start with what we already do and back in, get the theory from practice." Following this reasoning, the report offers a model for leader development:

The basic ingredient is good people. West Point takes great pains to admit young men and women who have demonstrated a readiness to learn, a willingness to take on responsibility. That's why the admissions committee looks for the above-average student who is also the team captain, a leader in her church, a volunteer firefighter.

Then there are four key elements of the developmental experience. The first—and West Point excels at this—is challenge: dragging cadets out of their comfort zone, giving them novel experiences and difficult goals, forcing them to resolve conflicts and take on new roles. There must also be a variety of challenges, from the physical to the purely intellectual. The goal is to make sure that no cadet can function solely in the arena he or she feels most comfortable in. Quiet cadets are made to speak up, the football players do gymnastics, the women take hand-to-hand combat.

To get the most out of these challenges, the cadets must have support, which is the second part of the model. Every member of the staff and faculty is a coach; professors and instructors are Army officers first. The third part of the model is assessment. USMA has a variety of feedback tools, some of them obvious: Cadets are graded for performance in leadership roles. Some of the assessment tools are not so obvious: A lot of self-examination goes on in the conversations among teammates, classmates, roommates. The fourth part of the model calls for reflection, for time to let the lessons sink in. Maturity doesn't come overnight.

The final part of the model is the freedom to fail. There is ample evidence in educational theory, Snook says, indicating that young people are most open to learning after they've experienced a failure, particularly one that challenges their assumptions.

I recognized parts of this model from my own cadet experience. The challenges were frequent, daunting, and often downright painful. Cadets have a saying that describes it aptly. West Point, they say, is, "a two-hundred-thousand-dollar education, shoved up your ass a nickel at a time."

I also remembered receiving and giving lots of coaching. During my years in the English Department, my boss was explicit about our duties: We were there to develop the next generation of military leaders, *and* to teach them to write clearly. I certainly remembered the assessment. Nearly every aspect of cadet life has some grade attached to it, down to the ridiculous, nit-picking detail.

But there were other parts of Snook's model I didn't recognize. There was little time for reflection and "the examined life." One former Superintendent, General William Westmoreland, said that the ideal West Pointer is a man of action—as opposed to a man of thought. Many at the Academy and in the Army equate reflection with touchy-feely, ivory-tower intellectualism at best; with navel-gazing egocentrism at worst.

Nor do I remember anyone telling me it was OK to fail. At West Point, as in many organizations, there is no room on the grade sheet for, "I dropped the ball, but I developed as a leader." In this highly quantified world, in the long, detailed record of their performance, cadets get little credit for trying and failing and learning.

Then there were the things Snook's model didn't address. I could remember no particular moment I could point to and say, "That's how they taught us about character." I also had questions about those cadets who are not served well by the West Point experience: those who flee at the first chance, and the ones who graduate, then cut themselves off from all contact with classmates. I know graduates who have been bitter for twenty years over things that happened to them in their first year as a cadet. Finally, as in any large group, there are those who just didn't get it, who remain dishonest, narrow-minded, bigoted.

Of course, my personal experience is dated. The only way to find out how today's West Point goes about its stated mission of building leaders of character was to go there, to see what happens or fails to happen. Because I am not versed in educational theory or psychology, I went about this the only way I know how: by looking for the stories. I followed a cross section of people through the course of a year, from the plebes at the bottom of the chain, to the Superintendent at the top.

During the nearly two years of researching and writing this book, I continued working as a keynote speaker, talking to business audiences about leadership. Time after time I met people who assume that military leadership has nothing to do with leadership in the civilian world. One businessman I know (who is not a veteran) characterized the military approach as, "You tell 'em what to do, and they have to do it, right?"

Not exactly. Sure, there is some room for autocratic leadership, the "do this or else" kind, but there are limits to what that can accomplish. On the other hand, there are almost no limits to what can be achieved by leaders who inspire people. In its most critical task—combat—the military practices an extreme form of decentralized leadership that makes today's dot-com wizards look like hidebound traditionalists. Current peace-keeping missions require an unprecedented degree of independent decision-making and flexibility. There are brand-new, one-year-out-of-college lieutenants on duty in Kosovo who are the de facto mayors of small towns. They act as mediators, judges, counselors, and police chiefs in villages torn by bloody strife and haunted by four-hundred-year-old vendettas.

Many of the young men and women who will take over those responsibilities in a year or two are at West Point. The ones who learn their lessons well will succeed in and out of uniform. This is the story of how they prepare.

DAY ONE:
WE'RE NOT IN
KANSAS ANYMORE

West Point, New York
June 29, 1998

A slim pamphlet published by West Point gives the following details about the United States Military Academy Class of 2002: Twelve thousand four hundred and forty applicant files were opened by the admissions office; 2,245 young men and women received congressional nominations (the first competitive hurdle) and met the academic and physical requirements of West Point. Twelve hundred and forty six were admitted.

Of these, 74 percent ranked in the top fifth of their high school class. None were in the bottom fifth. Sixty-four percent scored above 600 on the verbal portion of the SAT; 78 percent scored that well on the math portion. Two hundred and thirty-three received National Merit Scholar recognition, seventy-eight were valedictorians, 732 members of the National Honor Society; there were 224 Boys or Girls State delegates, 222 student body presidents, 191 editors or co-editors of school newspapers, 556 scouts. Of these, 139 were Eagle Scouts (men) or Gold Award winners (women). One thousand, one hundred and twenty-one of them—a whopping 89 percent—were varsity letter winners; 774 of them were team captains.

They are accomplished, educated, healthy, and willing to forgo much of what makes college life fun, including summer vacation. Today is their first day at West Point, and most of them are having trouble just walking and talking.

In the concrete and blacktop expanse called Central Area, a young man puts his left foot forward, on the command of the upperclass cadre member who is teaching drill. Inexplicably, his left arm swings forward. Since this eighteen-year-old learned to walk, probably around 1982, he's been doing it one way: left foot, right arm. The right foot comes out; the left arm does, too. Not today.

It's not that he isn't trying. His face is set, intense with concentration. He sweats, moves his lips as he repeats the commands. He doesn't look around, although he is a little disoriented. This day is meant to be disorienting.

"We want them to feel a little like Dorothy did when she landed in Oz and said, 'We're not in Kansas anymore, Toto,'" says Brigadier General John Abizaid, Commandant of Cadets.

Cadet Basic Training, also called CBT, also called "Beast Barracks" or simply "Beast," takes up most of the summer before freshman year. Six and a half weeks to learn how to look, walk and talk like soldiers; to begin to absorb—or be absorbed by—the military culture; to learn soldier skills, everything from how to march to how to fire a weapon; to learn how to obey.

There is a great deal to take in, and like so much of the West Point experience, it is accomplished pressure-cooker style. It shocks the delicate sensibilities of these teenagers who, for the most part, have led privileged lives in the wealthiest nation on Earth. It is this shock, as much as the fact that today is the beginning of the greatest adventure of their young lives, that makes R-Day memorable.

Four seniors—"firsties," in West Point jargon—stand on the low step outside Bradley Barracks, a six-story, L-shaped granite box that forms two towering sides of Central Area. Three of them are men; one of the men and the one woman are black. They wear the summer dress uniform called white over gray: white hat; pressed white shirt with gray epaulets and the black shield that marks them as seniors, or first class cadets; gray trousers with a black stripe running down the outside of each leg; leather shoes shined to a threatening luster. Each cadet also wears, as a badge of office, white gloves and a red sash that wraps around the waist. Thick tassels hang exactly over each cadet's right rear pants pocket.

This is "the cadet in the red sash," every West Pointer's first, unfriendly, welcoming committee.

A gaggle of new cadets lines up in four haphazard files. Green tape on the ground marks lanes, and they readily comply with the unspoken instruction to stand between the lines. At the top of each lane is the word "Stop," spelled out in the same green tape. Then a no-man's-land of a few feet and another line, behind which stands a burly senior wearing the red sash around his waist.

"New cadet," the firstie says in a voice meant for command. He raises one gloved hand, fingers extended to a knife-edge and aimed at the new cadet's nose.

"Step up to my line." He points at the line just inches from his gleaming shoes. "Not over my line or on my line but up to my line."

The new cadet steps forward, glances down, and aligns the toes of his shoes with the tape. The instructions come rapid-fire from the firstie, who punctuates every sentence with, "Do you understand, new cadet?"

No one pauses to acknowledge the moment, but something important has just taken place.

An hour ago most of the youngsters trying so hard to get to the

line . . . *not-on-the-line-or-over-the-line-but-to-the-line* . . . were
civilians, the majority of them just recent high school graduates. And
even if they didn't report to West Point with baggy jeans, exposed
boxer shorts, and skateboards, they were a lot closer to the denizens
of MTV than they were to soldiers.

Yet here they are, in the first few minutes of a career that will, for
some, last thirty years—and for others thirty hours—and not only are
they doing what they're told, they're trying to do it right. They are all,
to this point at least, willing participants in a long endeavor to turn
them into soldiers and leaders of soldiers.

A few of them may even be aware of the significance of this
moment. Many of them have spent months dreaming of the lofty
phrases of the admissions literature. They came, as one cadet wrote,
"for parades and rifles," dazzled by the name, by the history, by the
knowledge that they stand where many of America's great captains
stood. Others of them (and these will be the most unhappy) are here
because their parents want them to be here. For some, this is simply
the best school they could attend for free, or the only Division I
school to recruit them for sports. Under the gray sky they all look the
same: the ones who will become generals, and the ones who will drop
out in time to start classes at some other university.

"New cadet, you are allowed four responses: 'Yes, sir,' 'No, sir,'
'No excuse, sir,' and 'Sir, I do not understand.' " Then, with no pause,
the red-sash demands, "New cadet, what are your four responses?"

It takes a couple of tries before the neophytes learn the code. It
will take a little while longer for them to stop trying to explain things.
In that phrase, "No excuse, sir" (or "ma'am") is an early, critical lesson. Take responsibility for your actions. Always. No matter what the
consequences.

It is a lesson they will hear repeated for four years. Most of them
will get it.

The new cadets have been warned about the first day; some of them
by family members who have gone through this, some through careful attention to the recruiting literature, books, and documentaries.

They were even given helpful advice that morning at the official welcoming station, Michie (pronounced mike-ee) Stadium.

For most of the morning, a long line of candidates and their families stretches out behind the back gate of the football stadium. They enter in small groups, waved through a few hundred at a time by cadet ushers. They file in quietly, as if under some invisible instruction that this is a place of order, and sit in a section of the lower stadium seats. Before them, dressed in green "Class A" uniform of coat and tie, stands an Army colonel and a firstie.

"I'm Colonel Maureen LeBoeuf, head of the department of physical education."

LeBoeuf's official title, "Master of the Sword," dates from a time when West Point taught swordsmanship because it was a combat skill. She is five ten, with dark red hair cut short and stylish, the lean build of a runner. A few people in the crowd exchange appreciative glances; a couple of the fathers suck in their guts.

"Cadet Basic Training is both intensive and rigorous," she says in a gross understatement. "It requires dedication and motivation. I urge you to keep three things in mind during the coming weeks.

"First, remember to listen and do as you're told. If you're told to step up to the line," she says, turning her body so that she can take a long step on one of the bleacher seats, "step up to the line. Not over the line . . ." she takes a dainty step too far, "not short of the line, but UP TO THE LINE."

"Next, maintain a sense of humor."

"Third, remember that you are not alone. Every member of your class, every cadet before you, every member of the Long Gray Line has gone through this day. Experience tells us that it's best to take it one day at a time. With each day you will gain strength and confidence."

Although Colonel LeBoeuf is not a West Point graduate (she was already in college when West Point started admitting women), she is here in part because she is a model West Point would like all these young people to aspire to. A pioneer in Army aviation, a Ph.D. from the University of Georgia, the first woman to head a department here. Smart, successful, charismatic, with a sense of humor. *This is how it can turn out,* West Point says when she takes her spot this morning, with the sunlight glinting off the brass and silver and gold of her uniform. She is the same age as many of the younger parents in the stands, a perfect model for the *loco parentis* they all want to see.

LeBoeuf nods to the firstie beside her, who centers himself before the parents and candidates. Unlike LeBoeuf, who speaks naturally and from the heart, the cadet's speech is rehearsed, right down to the inflection. "West Point is a beautiful national landmark," he tells the families without a trace of enthusiasm.

And so it is. From the home bleachers one can see the low mountains across the Hudson River. There are long, unobstructed views up and down the valley, with Storm King Mountain rising to the left, New York City some forty miles downstream to the right. In fact, Michie Stadium was chosen by *Sports Illustrated* magazine as one of the most beautiful places in the world to view a sporting event. All over the bleachers, people crane their necks to take it all in.

"We urge the families here to enjoy your visit today."

Then he addresses the candidates directly. "At this time I'd like to ask the candidates to prepare to move down the stadium steps with your baggage."

He turns and indicates another cadet who has suddenly appeared far below, at the very bottom row of seats, beside a gate that leads from the stands and onto the football field. The distant cadet is at "parade rest," feet shoulder-width apart, head and eyes to the front, hands clasped in the small of his back.

"You will form a single file directly in front of the cadet you see standing there," he says, pointing. Then he turns back to the crowd. No more "please," no more "I'd like to ask . . ." This time it's just, "You have ninety seconds to say your good-byes."

A little ripple of shock rolls up the bleachers.

In the back row, Billie Wilson, a big football player from Texas, stands and palms his small bag. His little sister climbs up on the seat next to him so she can reach his neck for a hug. When her face appears above his shoulder, she bursts into tears. His mother's eyes are already red from crying, but she bites her lip to hold it together. Billie hugs his parents, shifts his bag to the other hand, and makes his way into the crowd moving down the bleachers.

The candidates start to line up. The first young man holds a guitar case in one hand, a suitcase in the other. The cadet at parade rest suddenly looks a little like Charon, preparing a boatload of souls to cross the river to Hades. When he is satisfied he has all that are coming, he turns smartly and steps out onto the playing field, leading them in a precise file across the fifty-yard line. A door opens in the opposite bleachers. Only one young woman in the line looks back over her shoulder. As they disappear under the visitors' stands, the families break into applause.

Maureen LeBoeuf appears again, standing next to the aisle as families file by on the way to the buses that will take them to their tour. Many of the parents thank her. One father chokes on, "Take care of my boy," and she says, "We will." When the younger brothers and sisters walk by, LeBoeuf frequently reaches out and touches them on the shoulder. One little boy of about ten, his face wet with tears, looks up at her.

"It's going to be all right," she says. "You'll see."

Many of them don't meet her eyes. Others try brave smiles. Some blink and squint as if in bright sunlight, although it is a cloudy day. On the other end of the home bleachers, another group is being processed by another colonel, another set of cadre members. Moments later another cadet, a woman, tells this group, "You will move out in ninety seconds."

Pete Haglin waited until the last possible minute to turn himself over.

"I was in the last group to go through at Michie Stadium," he says later. "I was so excited and nervous I don't remember much. I do remember walking across the fifty-yard line, and I could hear yelling coming out of the tunnel in front of us, but it was dark in there, and you couldn't really see what was going on. I wanted to look back, but I didn't."

Haglin has straight, almost-black hair inherited from his Korean-American mother; his height—about five eleven—comes from his father, a 1975 graduate of West Point. The elder Haglin, also named Peter, coached his son on what to expect on R-Day. He'd even made Pete practice reporting to the cadet in the red sash.

" 'Here's what you have to do,' he told me. So I knew. When they said, 'Drop your bag,' I dropped it. I didn't step on the line. It made things a little bit easier."

Haglin received his acceptance letter only weeks before R-Day. His parents had already made a deposit for housing at another college. Haglin knows the late notice means the admissions office had to work its way down the list of candidates before it got to his name. But none of that matters on R-Day. Haglin wants to be an artillery officer, like his father, so he takes a long-range view of the Academy: West Point is something to get through on his way to the "real" Army, the one he knows from his father's stories, and from his experience growing up an "Army brat" on posts all over the world.

The Haglin family—Pete, his parents, and two sisters—traveled together from Kansas City for R-Day. At the end of the briefing at Michie Stadium, when the ninety-second warning was given, his mother and sister dissolved into tears. But Pete was completely focused on what lay ahead.

Jacque Messel showed up at Michie Stadium by herself. She and her family had gotten the crying out at home.

"My parents said that it didn't make much sense for them to come along, that they couldn't really spend any time with me . . . but I think they were trying to make it easier on me," she says.

Messel is tall at five nine, with light brown hair and a résumé of clubs, honors, and athletics behind her. Her father is also a West Point graduate, class of 1968. On the night before R-Day, she stayed with the family of her father's classmate, a retired colonel who works at West Point. Jacque spent the evening watching television and trying to relax in someone else's home.

"He had to go to work early that day, so he drove me to the stadium. He stayed with me for a while, but then he left. Everyone around me was with their families."

Messel's father also tried to give her advice, but she wasn't as receptive as Pete Haglin. There were no practice reporting sessions or shoe-shining clinics.

"He did tell me it was all a big mental game," she says. "He was always big on teaching me responsibility and discipline. He'd make me get up in the morning and go running with him. This was in the summer, when all my friends were still sleeping in. And I always had jobs around the house, stuff I was responsible for."

Unlike Pete Haglin, who is headed into Beast willingly, Messel is reluctant, her commitment to West Point is not as strong. But once the acceptance letter came, and the family started talking it up and everyone started congratulating her, she felt like she couldn't back out.

On the morning of R-Day, Jacque Messel spent an hour and a half waiting in line at Michie Stadium, plenty of time for the anxiety to sink in. And it will never quite leave her, at least during basic training.

Bob Friesema remembers sleeping most of the way as he drove, with his parents and two younger brothers, from Wisconsin. The family spent part of the weekend before R-Day hiking at Bear Mountain State Park, which is just south of West Point. On Sunday, they went to church in the Cadet Chapel.

"It's this huge church, really impressive," says Friesema. "And it was good to know there was a nice church I could go to."

Church life has always been important to Friesema and his family, but the service wasn't exactly comforting. "The chaplain asked all the incoming new cadets to stand up, and everyone was looking at us. There were lots of cadre members there; I tried not to make eye contact with any of them."

Friesema spent his first hours of R-Day waiting in line outside the stadium.

"There were officers, admissions officers, I guess, going up and down and talking to families and candidates. They were being real nice, I guess so we'd know there were nice people in the Army."

Then came the shock of the ninety-second warning. "Mom started weeping right away. Then my little brothers started crying. I knew I had to get out of there before I lost it, too. I gave them a quick hug and left."

Friesema lined up quietly near the entrance to the field.

"I had my bag in my left hand," he says later, "because that's what everyone else was doing. We stepped out onto the field and there were these cadets walking along beside us. They started whispering under their breath, with their teeth clenched, saying things like 'Don't look around! Keep your head and eyes to the front! No talking!'

"And I thought, 'If they're doing this right out here on the field, where all the parents can still see us, I can't imagine what it's going to be like when we're out of sight.'"

By 11:00 in the morning the last new cadets have arrived from Michie Stadium. The large quadrangle of Central Area echoes with commands and martial music. Junior and senior cadets in white shirts and hats move about like border collies, shepherding new cadets here and there. Hundreds of new cadets are led in hurrying files back and forth to issue points in invisible basement rooms. They are issued big blue nylon bags full of supplies for their new life: underwear and socks and towels and shoe-shine equipment and boots and shoes and hats and gloves and belts and gym shorts. They are hurriedly fitted for gray trousers and white shirts. They are sent for quick haircuts and to drill stations to learn to march, to salute, to do facing movements. They are shepherded to lunch in waves; meals are measured in minutes. As the day wears on the big Alpha Company tote board (with the company motto "Aces Are Wild" across the top) fills up. The little white boxes beside the names of new cadets fill with check marks as they make their way through the stations and toward the parade. In between these stations, the new cadets check in with the cadet in the red sash.

"Sir, New Cadet Paley reports to the cadet in the red sash for the second time as ordered."

Paley needs three tries to get this right. She stumbles over the order of the words and is told to do it again because her voice has too much inflection.

"This isn't a conversation, this is a report," the red sash says.

Beside Paley, another new cadet renders a passable salute; his fingers tremble beside his eye. Behind her, other classmates move their lips silently as they practice the new language. All around them cadre members in white shirts bark orders. There is no yelling, but there is nothing pleasant about the sound or the experience. It is meant to be jarring, and it is.

The instructions come rapid-fire from the upperclass cadets, who end every sentence with, "Do you understand, new cadet?" No one speaks up or claims to not understand.

A large young man, his shirt soaked in sweat, moves his body with every word he speaks, as if he's using all the muscles of his chest to squeeze them out. "Hold still," a white shirt says. An upperclassman stands behind another new cadet and gives instructions. No inflection, no hint of human concern, just a rapid-fire string of words that is meant to impart information, but only to someone who can process things quickly. The new cadet keeps his head and eyes locked straight ahead. A tiny, nervous smile comes to his lips.

"Did I say something funny, new cadet?" the red sash snaps.

One new cadet is so tall that when he steps up to the line, he can't see the eyes of the cadet in the red sash, which are hidden beneath the black hat brim. He lets his eyes wander just as the red sash looks up.

"Is there a set of instructions for you on that wall behind me? Is there someone holding up a billboard to tell you what to say?"

By this time all the new cadets have been processed through a brief medical screening (one of many they have endured to get to this point). They have been issued a basic uniform of black socks, black gym shorts with gold letters that spell "ARMY," a gray T-shirt with the Academy crest, a plastic ID tag on a chain (worn around the neck), and a long paper tag with a list of in-processing stations to be checked off that day.

The new cadets don't look down at their own cards. The upper-class cadets check the list, mark the appropriate boxes ("HAIRCUT" or "UNIFORM ISSUE 1") and record each new cadet's progress on a large board that sits beside this barracks entrance. These new cadets are to join Alpha Company, but none of them know that yet. It's on the card, of course, but no one has given them permission to look down. With their paper tags dangling from white string, they look like a group of second-graders preparing for a class field trip.

"New cadet, are you wearing sunscreen?" a red sash asks in a tone that suggests there is some moral failure involved in being unprotected from the sun. The new cadet is sent to one of the green wooden tables nearby; a dozen bottles of Army-issue sunscreen and a stack of paper towels are piled there. He slops the cream on, rubbing it in with both hands on his freshly shaved head. Every few minutes the new cadets are sent to another table that holds metal canisters of cold water. During basic training the new cadets are constantly being told to drink water. Dehydration and heat injuries are preventable, and the chain of command is determined to ward off preventable injuries.

Other new cadets are sent to stand inside a rectangle taped on the pavement, where they wait in line. A thin second class—junior-cadet in white over gray strands nearby, collecting new cadets who haven't yet been to the barber shop. For the remainder of the summer, the new cadets will be escorted or travel in a group nearly everywhere they go. There is no sight-seeing, no strolling about the campus, no taking in the historical markers, no time to slow down and think.

"I will be moving very fast," the cadre member tells them. "You will keep up with me. You will keep your eyes on the back of the head in front of you."

These sentences could be placed alongside the Academy's official motto of Duty, Honor, Country. Everything here happens at the double-quick; there are always too many requirements and not enough time.

The new cadets stand in a tight file, feet spread shoulder-width apart, hands clasped in the small of the back. They do not talk, look around, or reach up to wipe sweat from their eyes. In the space of a few minutes, three different upperclass cadets ask, "Does anyone need to use the latrine?" They've been drinking ice-cold water by the cupful, yet no one raises a hand. It's unclear whether the new cadets know what a latrine is.

Twenty yards from the entrance to Bradley Barracks a young woman, a junior, teaches ten new cadets the fundamentals of drill. The woman wears her hair pulled up tight onto the back of her head. Firm voice, complete control of her material—she is all business. The young woman's mentor watches from twenty yards away. "She's ready to pass Drill Sergeant School right now," he says.

Sergeant First Class Tim Bingham, a combat engineer in the Army, is the tactical Non-Commissioned Officer—Tac NCO—for Alpha Company. Bingham, a stocky thirty-two-year-old, has thirteen years in the Army, including a three-year stint as a drill sergeant at Fort Leonard Wood, Missouri. His job is to teach the cadre—the upperclass cadets—about sergeants' work. In the Army, non-commissioned officers such as Bingham are the doers. Everything that gets done—from teaching a soldier to shoot to making sure a combat vehicle is ready to roll—gets done because some sergeant makes sure it gets done.

Bingham delights in bringing a healthy dose of Army-issue common sense to the intellectual development of cadets.

"Joe [a typical Army private] doesn't care if you've got a civil engineering degree," Bingham says. "He wants to know if you can take care of him . . . on a six-month deployment."

Looking around the crowded and noisy area with some delight, he notes how the cadre greet the new cadets. He summarizes it neatly as "Welcome to the team. This is how we do business."

Cadet basic training is run by cadets, juniors and seniors operating under the supervision of U.S. Army officers and non-commissioned officers. Alpha Company has two officers, two NCOs, and thirty-some cadet cadre. Their R-Day mission: turn the 158 civilians about to join Alpha Company into New Cadets—uniformed and marching in some semblance of order—in time for the Oath Ceremony scheduled for 4:00 that afternoon.

Cadet Kevin Bradley, a senior, is Alpha Company's twenty-year-old cadet commander. He got this job—a choice assignment—by demonstrating his leadership potential and earning high grades in military aptitude through his first three years. Bradley has the fit, scrubbed, and earnest look of many cadets: five nine, light brown hair cut close to his head, blue eyes. He tends to the quiet side.

"When he talks," an officer says of him, "the other cadets listen."

Bradley describes himself as "pretty much the standard West Point candidate," meaning he was no stranger to leadership positions even before he came to West Point. In his case that means captain of his high school football and baseball teams, student council president, and member of the National Honor Society. While the admissions office has no checklist of minimum achievements for a candidate, the academy does look for young men and women with "demonstrated leadership potential." Bradley was a good candidate for success on the day he walked in. In his three years at West Point he has worked his way up through positions of increasing responsibility, supervising anywhere from one to a handful of cadets. This summer is his biggest challenge.

Bradley will work most closely with Major Rob Olson, Alpha Company's Tactical Officer, or simply, "the Tac." Olson, West Point '87, has already spent ten years in the Army as an artillery officer. He is tall, loose-limbed, and talkative, given to jokes, Army aphorisms and the occasional profanity. He has an aw-shucks way of talking, as if what he's saying has just occurred to him and it might not be that important, but, gosh, since we're standing here . . .

Olson and Bradley spend much of the day watching. Their work, the preparation before R-Day, gives way to NCO work, which will be handled mostly by juniors. As he stands with the taller Olson, Bradley is learning an important lesson: how to stay out of his subordinates' way and let them do their jobs.

*   *   *

By noon the threatening sky has made good on its promise; it is raining. Not in the kind of drenching showers that sometimes visit here in
the summer, just an annoying drizzle. Some cadre members have
ordered their plebes to wear ponchos. None of the cadre members
wear rain gear because whatever keeps the rain out also traps so
much body heat that the wearer will certainly sweat through his
clothes.

Major Olson pulls Kevin Bradley aside; together they survey the
area.

"What's different here?" Olson asks. A few score of the hundreds
of new cadets around them are draped in the green plastic ponchos.

"Some of them are wearing ponchos," Bradley says. Anyone who's
been in the military more than a day knows it's called a uniform
because everyone is supposed to look the same.

That's the obvious answer, and Olson nods. But he expects more
from his cadet company commander.

"What's going on?" Olson prods.

"Well, some of the squad leaders decided their people should
wear ponchos," Bradley ventures.

This is also pretty obvious. The new cadets, part of the culture for
less than three hours, are already at a stage when they'll wear only
what they're told to wear. Besides, most of them wouldn't have known
to look for the poncho amid the piles of gear in their closets.

On the surface, this seems like a good idea. Common sense. The
Army issues rain gear so soldiers won't get wet when it rains.

"Right," Olson says. "Ordinarily, not a bad idea. What's wrong
with it today?"

Bradley is still thinking about uniformity, about appearances,
about how things will look for him and Alpha Company if some high-
ranking visitor comes along. He's still operating at the surface level.
Olson could simply tell Bradley what he's thinking, of course, but
Olson's job isn't to get Bradley to follow orders (Bradley already
knows how to do that.) He wants Bradley to think for himself, at a

level higher than "everybody needs to be in the same uniform." Olson wants Bradley to think like a commander.

"They're running up and down to the sixth floor," Bradley says.

"Right," Olson says. "Heatstroke city. So what do you think you should do?"

Bradley checks the sky. It isn't raining very hard; his own shoulders are just damp. "Let 'em get a little wet," he says with authority.

"You can always change the plan later," Olson reminds him. "You just gotta remember to have a plan."

Other seniors help Bradley run the company: He has an executive officer, a supply officer, and a training officer. Each of the four platoons has a senior platoon leader, a junior platoon sergeant, and a junior squad leader for each of the four squads. Seniors are first class cadets, or firsties; juniors are second class, also called "cows." (In the nineteenth century, cadets got only one furlough in four years: the entire summer between sophomore and junior year. Cadets left behind talked about school starting "when the cows come home.") Until the end of basic training, new arrivals are simply "new cadets." Those who make it through Beast will become plebes (as in plebeian, the Roman underclass), also called fourth class cadets.

Beast Barracks revolves around the squad leader and his or her relationship to the new cadets. The squad leader is the focal point of the new cadet's life: mother, father, counselor, coach, and trainer. It is a difficult, exhausting job, but it will teach the upperclass cadets who hold it the most about direct leadership. The cadets in this role say that their Beast squad leaders had a tremendous impact on them. Now they want to have the same effect on their new cadets.

Bradley and his cadre members have been here for almost two weeks before R-Day, rehearsing the canned speeches and practicing the skills the new cadets will learn during the first half of Beast Barracks. They talk about their own Beast experience like old soldiers who went through this fifteen or twenty years, instead of twenty-four months, earlier. Yet that short time has been enough to make them aware of the changes in West Point's recent history, in particular the move away from what is now considered "abusive" leadership.

"No one is going to be in their faces yelling at them when they make a mistake," a second class squad leader says. "But we're going to be right there making sure they do things right. That's harder on them and us."

"You don't want them doing things just because they're afraid to mess up," another squad leader says. "You want them to do things because they don't want to let people down."

This theme is repeated over and over again at West Point. The best leaders inspire their followers to *want* to perform. It is the party line. There is quite a bit of formal instruction on how to treat people with respect. There is even a senior whose job is "Brigade Respect for Others Officer," a fact that astonishes graduates who remember Beast and plebe year as a time of deliberate humiliation and even abuse. But today's cadets are not parroting the latest buzzwords. They have figured out on their own, through experimenting and being the subject of experiments, that leading is not about pushing.

By 2:00 the new cadets have moved into larger groups for drill. They move about the area in the blocklike formations they will use during the afternoon's parade. The drum beats steadily, *bah-boom-bah-boom-bah-boom,* the noise bouncing off the stone walls.

Inside the barracks, it is cooler, but not quieter as new cadets pound up and down the metal stairs. There is tape on the floor, running down the middle of the stairs and splitting each landing. The new cadets have been told to drive on the right side of the line, a precaution to help avoid collisions as they run up and down.

The rooms, small and Spartan, are designed for two, though most new cadet rooms will have three occupants. The linoleum floor is polished. The windows open on to Central Area, and the drumbeat and sharp commands carry up six stories. On one wall is a single bunk; the other wall has bunk beds. On each bunk are stacked a GI green blanket, sheets, a pillow. On the sink are three toothbrushes, toothbrush holders, bars of soap, plastic soap dishes, cans of shaving cream and disposable razors, all lined up like toy soldiers. Inside the open closet are rucksacks, helmets with camouflage covers, folded shelter halves, pistol belts, all the things that will remind the young men and women that they didn't choose Notre Dame or UCLA.

On the desk is a box of Academy stationery: heavy and formal with an engraved Academy crest. The new cadets will be required to write home tonight. There is a small paperback book with a camouflage cover, a manual of what the Army calls "Skill Level One" tasks: all the easiest things a soldier must master. There is also a copy of *Bugle Notes,* the "plebe bible," which is full of information and West Point lore, and a paperback study guide of fourth class knowledge. The new cadets will spend much of their summer memorizing information about the Academy (Who is the Dean? What are the names of the Army mules? What are the words to the Alma Mater?) and the Army's equipment and organization.

Also on the desk is a one-page letter written by the squad leader for the room's occupants. Though it appears to have been written in a rush (and would make the squad leader's English professor cringe), it carries a critical message for the new cadet.

WELCOME!!! You are now a member of the Good Ole Gang, 4th squad 3rd platoon. I Cadet Jett will be your squad leader for the next three weeks. You have just started to infringe on the beginning of your West Point experience today, 29 June 1998. I as well as the rest of your company cadre (The Aces) are looking forward to working with you this summer. You will experience many new things in your life this summer as well as the future that lies ahead in your road to success at USMA. There will be times that will seem impossible this summer and others that will be more fun. No matter what you are faced with you must stay motivated and work as a cohesive squad to make it through. You will hear me speak of team work quite often the next three weeks, because it will be the key reason for your successes and failures not only this summer, but the rest of your future at West Point.

The author, squad leader Grady Jett, is a wide receiver on the Army football team. He believes in attitude, practice, and teamwork, and his squad will hear those words over and over in the coming weeks. The upbeat tone of Jett's letter contrasts with a famous "Welcome" speech delivered to a long line of plebe classes by a now-retired department head.

"Look to your left and right," he said. "One of the three of you won't be here on graduation day."

For the first two years, cadets have the option of resigning, with no commitment. They can leave free and clear and owe the government nothing for the two-year scholarship. The decision point is the first academic class of junior year. A cadet who is present at the beginning of that class is obligated to serve for five years after graduation. A cadet dismissed during junior or senior year (for violations of the honor code, academic failures, or as part of a disciplinary action) can be sent into the field Army as a private soldier.

While at West Point, cadets are members of the United States Army, subject to military law and discipline. They are also on full scholarship, which includes tuition, room, board, medical and dental care, and a stipend of seventy-two hundred dollars a year, most of which goes to paying for books, computers, and uniforms.

Downstairs, a squad of new cadets is lined up in a single rank facing their squad leader, a young woman who wears her hat pulled low. She leans her head back slightly to peer from beneath the brim. Her charges, who were civilians when they ate breakfast this morning, stand in a straight, evenly spaced rank. They are all dressed in white shirts, gray cadet trousers, and black shoes. (They were told to bring well-broken-in, plain-toe black leather shoes from home. The few who didn't bother to break-in the shoes already regret it.) The men have fresh, severe haircuts; the women wear their hair very short or pulled up into tight buns.

"What are your four responses?" the squad leader demands.

All around them other groups are drilling; there are overlapping marching commands, the heavy beat of the bass drum, now accompanied by several bugles. Other squads and individuals are reciting their four responses, or answering the constant question, "Do you understand?" But this group is intently focused on the woman in front of them. They have known of her existence for only a few intense hours. They don't know her first name, or where she's from, or much about what she expects of them. Most don't know what she can do for them, or even that she has been preparing for this role for weeks, months. Most of them cannot imagine that this apparition of military exactness, with the sharp uniform and erect carriage and command voice, was in their position just two years earlier. They know none of this because no one has told them, because no one thinks it's important they know anything other than the four responses. And everything about them is focused on answering her correctly. They respond in perfect unison, rattling the windows, oblivious to everything but the need to perform.

"Ma'am, my four responses are . . ."

Captain Brian Turner is the associate tactical officer of Alpha Company. A 1991 graduate of West Point, Turner just finished a year in the Tactical Officer Education Program (TOEP, rhymes with rope). At the end of the summer, he will become the Tac of Company F-2, one of the academic year companies. He is twenty-nine years old, unmarried. During Beast, he is an assistant and understudy to the more experienced Rob Olson. Turner and Olson could hardly appear less alike—except for their uniforms.

Rob Olson is a tall, thin, and white. Raised in Minnesota, he played one season of hockey at West Point. He is a natural storyteller, full of self-confidence, a man who gives the impression that nothing would rattle him. He has been tested in a variety of field assignments and in Desert Storm. Based on this performance, Olson was promoted ahead of his peers; he pinned on the gold oak leaves of a major a year before most of his classmates. Rob is married to a West Point classmate (now an Army doctor) and is the father of two small children.

Brian Turner is black and shorter than Olson, with the compact muscularity of a wrestler. In the minimalist language of stage directions, he would be described as eager. Like many of the young officers and most of the cadets, Turner is all energy. He can barely stand still, but shifts his weight back and forth from one foot to the other. In conversation, he swings back and forth from the very serious to a playful broad grin. Five eight, with coffee-dark skin, a native of inner-city Chicago, he is losing his hair prematurely, a look he helps along by keeping his head close-shaved. When he removes his hat for a photo he jokes, "I have so much forehead they call me five head."

Tuner, like Olson, has commanded troops. He is eager to marry his army experience with the theories he has learned in the TOEP program. He speech is rapid-fire, full of Army jargon.

"The cadre has to plan, resource, train," he says. "Cadets who don't make the standard this summer don't get a spring break, so the plebes will be motivated to pass their skills tests."

"The focus of the summer is on the cadre. If the new cadets fail, it's on the leadership. If the platoon is unable to pass, we'll look at the platoon leader. This is a steep learning curve for these cadets over the summer."

Turner walks up the six flights of stairs to where the new cadets of Alpha Company are stowing gear, sorting through uniforms. As the day wears on, the new cadets undergo a visible transformation, going from gym shorts and athletic T-shirts to gray trousers and athletic shirts, to gray trousers and the white uniform shirts they will wear in the Oath Ceremony on the Plain this afternoon.

"Somewhere in this hallway is a future general," Turner says. "They're definitely Generation X, though. They want to know the 'why' behind everything. They're sponges, they want to learn about leadership."

He acknowledges that changes have been made in the way basic training is run.

"Beast is just as hard as it ever was," Turner says confidently. "Now it's hard on everyone. I mean, the old grads have been bad-mouthing changes of any kind since MacArthur reformed the place."

When Douglas MacArthur returned to West Point from the battlefields of World War I, he was a decorated combat veteran, cited for gallantry. He entertained the idea—considered outlandish by many— that cadets should not spend their summers on the parade field, as they had throughout the nineteenth century, but should have to learn the combat skills required of junior officers. He ran into stiff opposition, especially from the academic board, the tenured professors who saw it as their job to keep West Point as it has always been. The battle is part of Academy lore.

In recent years West Point has undergone another period of major change in its attempt to stay a top-ranked college and provide leaders for a rapidly changing Army. But even while administrators update the curriculum, seek funds for the physical plant, study the demographics of college applicant populations, solicit input from the Army on the quality of graduates, and track the promotions and separations of West Pointers, the alumni (known collectively as "old grads") mostly seem interested in whether or not plebes are getting yelled at. The most pitched battles have been over attempts to get rid of pointless abuse while maintaining a tough—and toughening— experience.

The Class of 2002 will remember R-Day as a visceral experience; nevertheless, each succeeding class thinks the ones following have it easier. And in fact the expression, "The corps has . . ." (as in "the corps has gone to hell") is so well known and accepted as true that it needn't even be finished. Use the first three words and a West Point graduate will not only get the message, he or she will probably agree.

Some alumni recognize the usefulness of their own experience— even if they hated it—while accepting that change might be good.

The old plebe experience paid off for John Calabro, '68. "When I was a kid I was a wimpy evader of difficult things. This place broke me of that. When I came up here I wasn't sure what to expect. The first stop was at the gym; we were just shoved around by some disgruntled enlisted men and I thought to myself, 'I can handle this.' Then a soldier marched us to the edge of Central Area and pointed across this vast expanse of open space to a cadet in a red sash and said, 'That's where you're going.' "

"As soon as we hit that area, all hell broke loose, and I thought to myself, 'This is what they were talking about.' "

"What they were talking about" was the noise; the screaming; the nose-to-nose, spittle-flying screeching of upper class into the faces and ears of shocked new cadets. It was a hallmark of plebe year that started on R-Day and lasted until graduation: eleven months of hell.

It's gone.

Brigadier General John Abizaid, '73, is determined to make West Point more like the Army and less like initiation into some fraternity.

"Here's the way someone explained [the changes] to me," Calabro says. "You have the trainee, the trainer, and the task. It used to be that the trainer was part of the problem, part of what was making the task difficult. Now the idea is that the trainer is part of the solution. If you want to make things difficult, you raise the standards of the task, make the task harder."

This is a move toward what the academy calls "inspirational leadership." It is also more in line with the paternalistic nature of Army leadership than was the old system.

In his novel *Honor and Duty*, Gus Lee, a 1968 classmate of John Calabro's, begins his story by describing the blast-furnace quality of the first day. Lee details the abuse handed out by upperclass cadets, twenty- and twenty-one-year-old youths with near-absolute power. Everything about the plebe's life was at the whim of whatever upperclass cadet happened to be around. As late as the seventies plebes were ordered to "pass out their plates," returning untouched meals to the waiter for an error in the bizarre table etiquette of the Cadet Mess, or for incurring the displeasure of an upperclass cadet. Lee writes about eating toothpaste in his room, of smuggling little plastic packets of jelly out of the Mess Hall and licking them clean in the darkness after taps.

Yet some treasure the system that treated them this way, not for the treatment itself, but for the results it produced.

Charles Murray, USMA '62, now practices law in Florida after retiring as a colonel. Murray is almost wistful as he speaks of the hazing that helped shape him. There were shower formations, in which plebes dressed in sweat suits and raincoats were made to stand in steamy shower rooms until they nearly fainted; uniform drills in which plebes were given orders to appear in a certain uniform in an impossibly short amount of time, the game repeated through endless varieties of uniforms until the plebes dropped from exhaustion in rooms littered with articles of clothing and equipment. And then, almost inevitably, an upperclassman would announce "Room inspection in ten minutes!"

"Steel is forged in fire," Murray says. He is worried that the furnace has cooled too much. He admits that some of the hazing was just stupid, boys'-school and fraternity-row stuff, but he also insists it was useful in the ultimate test: combat.

In Vietnam in August 1966, Murray led a relief force that fought its way to a besieged company of American infantry. The company commander had been killed; Murray took command of the shot-up unit in the middle of a firefight, surrounded, he says, "like Custer."

"I had bodies of my own men, including my classmate [the dead commander], piled around me. We could hear the enemy shouting and hollering, like they were holding a football pep rally, all through the night," Murray says with a little laugh. "I thought it was going to be the shortest command in history."

When he talks of that night, with his new command decimated by fire, with dead and wounded all around him and the enemy within earshot massing for an attack, Murray credits plebe year with giving him what he needed to perform.

"Plebe year is supposed to teach you how to function under pressure, how to control your emotions and still make decisions when people are counting on you. I'm not sure plebe year does that anymore."

In February 1998, hundreds of West Pointers thrilled at the following e-mail message posted by Bo Friesen, '83. It was passed around and around among the classes with the recommendation, "Here's an awesome defense of the old 4th class system."

As I went through [plebe year], I did not understand how cutting a cake into nine equal pieces [plebes were required to prepare dessert servings] would help an officer lead soldiers into battle. The myriad of disjointed memorizations, ludicrous tasks and perpetual panic mode seemed to have very little to do with the profession of arms. I maintained this attitude throughout my upperclass years . . . right up until the moment I commanded a cavalry troop in the Gulf War.

One night, around 0100, we conducted a passage of lines to assault an airfield. We had gone almost 60 hours without sleep and it was raining with a vengeance (yes, rain in the desert . . . lots of it). Our own artillery was falling short and landing amongst us, one of my platoon leaders was heading off in a tangent to the direction he should have been following, the squadron main body was drifting too far to the north, my driver was heading straight for a ravine, a tank in my 4th platoon threw a track, we found ourselves in the middle of one of our own . . . minefields, the objective was spotted on our right flank (instead of in front of us, where it should have been), almost no maps existed for our area of operations, my boss was perpetually screaming for me to change to his [radio] frequency (an impossibility with the wonderfully designed, single-transmitter command tanks), a half dozen spot reports were coming in from my troops (all critical), my intell NCO had a critical update, my XO [executive officer] had a critical update, my ops NCO had a critical update, my 1SG [first sergeant] had a critical update, my gunner had spotted dismounts [enemy soldiers on foot]. The regimental commander was forward with us adding his own personal guidance, visibility was almost zero, there was a suspected use of chemical weapons, regimental S-2 [intelligence] reported 500 heavily armed Republican Guards on our objective (later determined to be a squad of American engineers), and I had a moderate to severe case of dysentery. (A run-on sentence, I know, but then again it was a run-on night.)

It was during this little slice of heaven (of all places) that the 4th Class System was illuminated to me in all its glory. Its goal was not harassment, ridicule or punishment. Its goal was to train the neural network to deal with an overwhelming amount of disjointed information, quickly process that information, categorize it, and make rapid, sound decisions. At that moment, I would have gladly given a month's pay to the genius who devised the 4th Class System.

Graduates who defend what the old system did for plebes never add, "and it taught good leadership techniques to the upperclass cadets." West Pointers do not say that learning how to abuse, insult, and intimidate subordinates proved useful later on. And it is on those grounds—that the old system taught bad habits to upperclass cadets—that the system was attacked.

At an alumni luncheon at the West Point Officers' Mess, a member of the class of 1970 told how he had taken the lessons he'd learned as an upperclass cadet to the Army, with almost disastrous results.

"We were in Berlin, and so we had pretty good soldiers. One day I heard that one of my men had dope in the barracks. I went to his room and started freaking out, turned over his locker, spilled all his clothes, flipped over his bunk. I was treating him like a plebe. And my platoon sergeant grabbed me by the arm and pulled me out of there and said, 'Sir, that shit might work at West Point, but it ain't gonna cut it here.'"

The storyteller was embarrassed, almost thirty years after the incident. "I had to completely relearn how to deal with people after I left [West Point]," he said.

The effort to change the relationships between leader and led have been slow in coming and resisted as much by cadets as by alumni. But the changes have come. The Cadet Leader Development System, introduced in the eighties, overhauled the way cadets live and work together. CLDS (pronounced "cleds") informs almost everything about the cadet experience at West Point; it is about developing leaders.

The system was designed to ensure that cadets are given

increased responsibilities over their four years. They move up through positions of greater responsibility and authority; they are exposed to the Regular Army through a variety of summer programs. A former Superintendent, Lieutenant General Dave Palmer, liked to say that West Point went from the fourth class development system (the old plebe system) to a "four class development system." The culture changed from one in which plebes were held to high standards while everyone else slacked off, to one in which the upper three classes are held to increasingly higher standards as they progress through four years.

"If we let your son in here, we assume he can graduate," Colonel Bob Johnson tells the parents of recruited football players at a reception the night before R-Day, 1998. "That's our job. We're surrogate parents. I'm not going to hug him or kiss him, but I will kick him in the butt when he needs it. I'm going to get eyeball to eyeball with any cadet who's having trouble with academics and I'm going to tell him to get his act together."

Johnson, a former Army football player, looks like some casting director's choice for an infantry colonel. He is dressed completely in black, with clothes that show his athlete's physique, and a shaved head that makes him look a little like a black Mr. Clean.

"We're here to build leaders, not people who call home crying every five minutes. We will be fair, and we will be hard. We will teach him to be responsible, we will teach him to solve problems; when he gets home you will see a change in his demeanor."

"We'll keep them straight and we'll send them back to you better than you sent them to us."

R-Day speeds on. In Central Area the drumbeat pounds through another hour. The new cadets now march in larger units, platoons and companies of 150. They may not be ready to join the whole corps on parade, but they can do a fair imitation of a mediocre high school drill team.

Cadre members scurry around the formation, checking to ensure that every new cadet has the right uniform. All 158 new cadets assigned to Alpha Company have been screened, measured, fitted, shorn, checked and rechecked. Each one is now under the control of a squad leader; the squad leaders answer to the platoon leader/platoon sergeant team, who in turn answer to the company commander, Kevin Bradley. Bradley answers to Olson for every body, every piece of equipment.

Naturally, not everything goes right. Some new cadets are missing bedding. There are prescription eyeglasses to be picked up (the Army-issue frames that look like bug eyes; no stylish civilian glasses after today). The big clock in Central Area moves toward parade time.

Bradley talks aloud about competing demands. Squad leaders don't want to come outside too early and wind up standing around while other squads are still upstairs; the minutes are precious and too few. They'd rather use the time in the barracks to get things set up for the next day's training. There are a staggering number of details squad leaders must teach these civilians, these almost-new-cadets: everything from how to make a bed to how to stow their uniforms to where the latrines and showers are. This means that each squad leader will hang back in the barracks until the last possible moment, trying to wring out of the clock just another minute of time to get things ready for the next requirement. And there is always a next requirement.

Major Rob Olson and Captain Brian Turner do not interfere. They are there to make sure that the cadets can recover from any minor mistake, and that they don't blow any of the big things.

A platoon leader whispers something to Bradley. A new cadet has urinated on himself, probably a victim of "drink more water, DRINK MORE WATER," and the decidedly unfriendly way the cadre asks, "Who has to use the latrine?"

The platoon sergeant handled it smoothly, hustling the youngster upstairs where, fortunately, there is another pair of gray trousers that fit. No one laughs, at least not in the presence of the new cadet.

"What's your most important mission?" Olson asks Bradley as the cadet looks around the area, counts the squads ready for company drill, and figures out how many are still upstairs.

"Sir, the most important thing is to get everyone out there on time for the parade," Bradley says.

"And they all gotta know how to march," Olson adds.

Bradley smiles. "Yes, sir."

"Keep focused on that. Let the squad leaders have as much time as you can afford to give them—which may not be as much as they want."

Kevin Bradley gets the distinction, and moments later the 158 new cadets of Alpha Company are forming for the last block of drill instruction before their first ceremony on West Point's famous parade ground, called "the Plain."

Meanwhile, the families of new cadets wait near the entrances to Central Area, which are closed off by heavy chains. At the corner nearest the old First Division barracks, a hundred or so family members sit and stand, peering into the area and trying to catch a glimpse of their new cadet. Many of them use binoculars. A couple of teenage girls in shorts sit cross-legged on the ground, their shoulders slumped forward like tired sentries. Suddenly one of them straightens, points, and yells, "There she is!" She is pointing at what looks like an undifferentiated mass of gray. With their new haircuts, new clothes, new way of walking, it would be difficult even for a family member to pick out one new cadet among the hundreds visible at a distance. The missing sister, even if she were only a few yards away, is no longer allowed to turn and wave, so the family has to be content with the unconfirmed sighting.

A hundred yards away, the inside of the Cadet Library is cool and quiet. Just inside the reference room is a long display case, twelve feet of polished cherry with a glass front. In the center, a wooden plaque says "General of the Army" the gold lettering is flanked with five stars in a pentagon pattern. Displayed there are the class rings of three of the West Pointers who wore five stars: Douglas MacArthur, 1903; Omar Bradley and Dwight Eisenhower, both 1915.

The rest of the shelves are a display of class rings dating back to the mid-nineteenth century. It's part museum piece (West Point claims to be the first undergraduate institution to give class rings); and part something else: a memorial, a *memento mori*, a reminder, to any cadet who might stop and read the tiny inscriptions, of the Academy's ultimate mission.

[Class of] 1988
ILT [First Lieutenant] Donaldson Preston Tillar III
Killed in action in the Persian Gulf War, 27 February 1991

1969
CPT [Captain] Paul Coburn Sawtelle
Killed in Action in Vietnam
16 April 1971

January 1943
ILT George Eberle
Killed in action at Normandy
6 June 1944

1940
CPT Joseph V. Iacobucci
Died in Prison Camp Fukuoka, Japan
14 March 1945

1841
MG [Major General] Amiel Weeks Whipple
Mortally wounded at Chancellorsville [Virginia]
7 May 1863
Gift of his grandson COL [Colonel] Sherburne Whipple

1841
LTC [Lieutenant Colonel] Julius P. Garesche
Killed in action at Stone River
31 December 1862

1850
BG CSA [Brigadier General, Confederate States Army]
Armistead L. Long

1892
ILT Dennis M. Michie
Killed in action at San Juan River, Cuba
1 July 1898

There are only a few librarians around. The reference room is quiet. Outside, the clouds are breaking up over the parade field. The library's big arched windows frame the statue of General George S. Patton, 1909, which stands directly across the street. A dozen people in civilian clothes, in shorts and sensible shoes, circle the general, snap his picture, or pose in front of the statue's base.

At 4:00 the parade is about to start. The Plain is a flat patch of emerald on a bluff above the river. One side is braced by the long granite wings of Eisenhower and MacArthur Barracks. Opposite are aluminum bleachers, the massed cannons and tall trees of Trophy Point, then the long valley and flat silver plate of the Hudson River.

The cadet companies pour out of the sally ports when the band strikes the first chords. The cadre members are dressed in white: starched trousers and high-collared tunics. The new cadets wear short-sleeved white shirts and gray trousers, white gloves, and no hats. This is the only parade, for their entire four years, in which their heads will be uncovered. (There is not enough time in the day's tight schedule to fit everyone for hats.)

From the stands, from a distance, their movements are surprisingly smooth. They salute, stand at parade rest, then back to attention on command. Brigadier General John Abizaid administers the oath, his voice echoing against the barracks and the green hills beyond.

I, ___, do solemnly swear that I will support the Constitution of the United States, and bear true allegiance to the

National Government; that I will maintain and defend the sovereignty of the United States, paramount to and all allegiance, sovereignty, or fealty I may owe to any State or country whatsoever; and that I will at all times obey the legal orders of my superior officers, and the Uniform Code of Military Justice.

For those families who came here thinking that West Point was just another college on a list of schools their talented sons and daughters might have attended, the sight of all those uniforms and the solemnity of the Oath Ceremony reminds them that this is something different. They stand on the bleachers to take in the sight, and the spectacle is impressive. This morning's thirteen hundred civilians have been changed dramatically, and if they're not yet soldiers, they are no longer high school students.

Up in the aluminum stands, two women who look remarkably alike, perhaps the mother and grandmother of a new cadet, sing with gusto when the band plays the National Anthem. Across the whole of the bleachers, there is a good deal more enthusiasm for the song than is usually heard at baseball games. There is a flurry of commands and sharp salutes, though the flags stir only occasionally in the heavy summer air. Then the companies pass in review, wheeling by the Superintendent (the three-star general who is both president of the college and military commander of West Point) and the Commandant (the one-star who is responsible for cadets' military training). There are ripples of excitement as the cadets pass close to the bleachers, while families strain to spot a son or daughter. The new cadets tramp by at 120 steps per minute, on past the generals, on past their families; they wheel sharply in front of the white confection that is the Superintendent's House, then, quickly, the class of 2002 disappears back into the dark tunnels.

For the mothers and fathers, sisters, brothers, and grandparents, the parade marks the end of something momentous. They will leave West Point without the children who have been with them for seventeen or eighteen years. Understandably, the families take their time getting out of the stands, as if they're unsure that their part in the drama is over.

In the barracks, the members of West Point's Class of 2002 have hours of unfamiliar tasks ahead of them. There is equipment to draw and store, rules to be memorized, whole new ways of speaking and eating and walking to be mastered. Somewhere in the course of the evening, they will be told to sit down and write a letter home. Most of them will manage only a line or two, something about all the work to be done on this, the first of many long days ahead of them.

BEAST

The first days of Beast Barracks slide by in a blur for the new cadets. If they are remarkable for anything, it is for the mundane and seemingly endless nature of the tasks to be learned: how to wear various uniforms, how to salute, eat in the Mess Hall, march, carry a rifle. But over the first week, the details add up, and by the time the new cadets climb down off the trucks for their first day of field training, they have started to look like GIs.

They clamber out of the big five-ton trucks onto the dusty road wearing battle dress uniform, or BDUs, the familiar baggy camouflage. They carry big green rucksacks, with rolled foam pads slung across the top; load-bearing equipment (called LBE); a set of suspenders and a belt from which hang canteens, first aid, and ammunition pouches. They wear the coal-bucket Kevlar helmet, which everyone calls a K-pot, and carry the M16A2 rifle, standard issue for the Army and Marine Corps. In the past few days they have learned to put all of this together: the hooks and buckles and belts and straps, so that they at least look like soldiers. They have learned how to form into squads and platoons, to respond to marching commands to get them from one place to another. They stand straight, head and eyes to the front, and wait until they're told to move.

This first day of field training has brought them to a pleasant, shady hillside on West Point's sprawling reservation, the "NBC Site," where they will be introduced to Nuclear, Biological and Chemical Warfare. They will learn how to wear the "MOPP" suit, a charcoal-treated coat and pants that, with its rubber gloves and overshoes, is supposed to keep the soldier safe from chemical weapons. They will learn to wear the protective masks, the huge, black-rubber headpiece with its monstrous goggle-eyes and a green plastic hood. Wearing it is like wearing a head-sized, portable sauna.

The biggest challenges for the new cadets are simple tasks: learning to wear their specialized equipment, for instance. Upperclass cadets like junior Greg Stitt are on to more difficult things: Stitt is learning to be a platoon sergeant. Just as the squad leaders take care of their ten or eleven new cadets, Greg Stitt takes care of four squads, seeing to all the details that make the machine run smoothly: food, water, transportation, accountability. He gives instructions to the squad leaders, passes information from the platoon leader—a senior—and generally acts as a second-in-command and chief of operations.

Stitt stands in a central location, in the same uniform as the new cadets, including the weapon. As they complete their various tasks, the new cadets report to Stitt, who records their successes on a card. New cadets who don't complete the summer's required training will lose their spring leave and take the tests over again while the rest of the class enjoys the break. This is another innovation of the Commandant, Brigadier General John Abizaid: Rewards are tied to performance, just as in the regular Army, where a soldier who hasn't mastered the required skills is not promoted.

Stitt, who was an Army enlisted man before coming to West Point, thinks these are good changes.

"The Comm has some good ideas; he's moving this place to be more like the army."

Stitt looks up as a new cadet reports that he has successfully completed the task, "Don the protective mask." Stitt nods, makes a quick pencil mark, and tells the new cadet, "Move out."

Greg Stitt isn't losing any sleep worrying whether or not CBT is too easy, or if the new cadets should be yelled at more. He thinks new cadets should be treated like basic trainees in the Army. That way, he says, "They'll know what it's like to be a private, and they'll learn how to treat other people."

Stitt, who is older than his classmates, is five eight, with red hair and the compact build and coiled energy of a lightweight boxer. He was a helicopter crew chief in the 82nd Airborne Division before applying for West Point. His experience as an enlisted soldier has shaped his view of how leaders should treat subordinates.

"Most people don't realize they're being developed until after it's over," he says. "If they pay attention, they realize they can learn from both good and bad examples. I can learn good points, or I can learn bad points," he notes. "There is a lot of emphasis on thinking for yourself."

All through a hot afternoon, the new cadets move from one station to another in squad groups. The equipment is unfamiliar (except to those new cadets who were enlisted soldiers and thus have been through Army Basic Training). Every station includes some timed task, but the pressure is not great; the tasks are just not that tough for this bunch of two-varsity-letter-winner, National-Honor-Society new cadets. A letter home might sum it all up as "spent the day getting dressed and undressed in unfamiliar and uncomfortable extra clothing." The new cadets joke among themselves, compare notes on hometowns, slouch, and stretch under the shade of the trees as the classes drone on.

Major Rob Olson, Alpha Company's Tac, notes that this is the first time the new cadets have had the chance to talk to one another at length, to find out they're not the only ones worried about fitting in, about handling the stresses of basic training. "For some of them this is the first time they've smiled [in the seven days since R-Day]. And there was a line at the latrine. It's the first time some of them have taken a shit in a week."

The new cadets stand in clusters in the shade after removing their equipment. Squad leader Grady Jett, an Army football player from Houston with a TV-star cleft in his chin, lets the new cadets joke around a bit, just a little banter, but he doesn't hesitate to put them in the front-leaning rest—the push-up position—if they don't respond fast enough to instructions.

Platoon Sergeant Stitt, who stands off to the side, says the platoon sergeants and company first sergeant—the highest-ranking cadet NCOs in the company—are saddled with a great deal of sudden, wide responsibility, especially considering that for most of them their entire leadership experience has been supervising one or two plebes during the previous semester. Now they're responsible for forty or fifty, or in the case of Company First Sergeant Josh Gilliam, 158 new cadets.

"We're allowed to try things and do things differently," Stitt says. "They [the Tacs] guide us into the right lane." He holds up his hands to indicate a left and right limit. This is the same description Olson used when talking about letting the cadre figure things out on their own.

A new cadet appears beside Stitt to report that he has successfully completed one of the required tasks. Perhaps inspired by the casual surroundings, the new cadet stands in a relaxed posture. Stitt, who has a serious demeanor for a young man, doesn't even vary his voice as he says, "What do you want?"

"I completed . . ."

"Don't be dropping 'sirs' or I'll be dropping you," Stitt interrupts.

"Sir, I have completed the task, 'don protective mask.' "

Stitt makes a note on a card that lists the names of all forty-plus new cadets in his charge. A moment later another new cadet walks by on his way to the Porta-John that serves the site. He is bareheaded.

"Where's your helmet?" Stitt asks.

The new cadet looks at the platoon sergeant, then scrambles to retrieve the helmet, which he jams onto his head.

"Drop," Stitt says, matter-of-factly.

The new cadet, who is also carrying his rifle, bends over to get into the push-up position, but he doesn't know what to do with the weapon. Somehow he knows he shouldn't just lay it on the ground.

"Like this," Stitt says. He drops into the front-leaning rest, his rifle resting on the backs of his hands.

Stitt didn't get to West Point by the traditional route. The son of an Air Force enlisted man and grandson of a World War II Army Air Forces veteran, Stitt enlisted to become a helicopter pilot. He ended up at Fort Bragg, flying in the back of the Army's workhorse aircraft, the Blackhawk.

His lieutenant saw his potential and said to him, "Hey, Stitt, you're not married, you've got good SATs; how about applying for West Point?"

"I told him, 'No way, sir. I'm not a school guy,' " Stitt says, narrating.

"And then one day I was on detail at the [82nd Airborne Division] museum. I'm sweeping up and I notice a display about General [Jim] Gavin."

Gavin, USMA '29, commanded the 505th Parachute Infantry regiment on D Day, and later commanded the entire division. He is a legend in a unit with no shortage of heroes. Even as a general, he carried a rifle into battle. In today's division headquarters at Fort Bragg, North Carolina, an entire wall is covered with photographs of division commanders; Gavin is the only one wearing a helmet in his photo.

"I looked in this case and I noticed that General Gavin was an enlisted man before he came to West Point," Stitt says. "So I went back to the lieutenant and said I'd give it a try. I wrote the essays [for admission] the week they were due. I did everything at the last minute."

Stitt's claim to fame as a cadet is an attempted "spirit mission" just before the Army-Navy football game the previous fall.

"Spirit mission" is a catchall description for almost any high-spirited, unorthodox activity that is designed to shake up the status quo; organizers of the most imaginative missions can taste a bit of fame. Douglas MacArthur, who was number-one man in his class of 1903, engineered the removal of the reveille cannon to the top of the clock tower in what is now Pershing Barracks. It took the post engineers, working in daylight, several days to remove the cannon the cadets had moved in one night. Other cadets have won a degree of notoriety, and many of them have been punished heavily when caught, for such antics as filling the Commandant's office from floor to ceiling with balled-up newspapers; for dragging cannons from Trophy Point to various places on post, such as in front of and aimed at the Superintendent's quarters. A favorite of modern times is the kidnapping of the Naval Academy mascot, Bill the Goat.

"I'm the only cadet, as far as I know, who has ever climbed Battle Monument and kissed Victory," Stitt says without smiling.

Battle Monument, a giant granite shaft that sits on Trophy Point, is topped with a statue of Winged Victory that sits some sixty feet above the ground. It is a memorial to the officers and men of the Regular Army who died in the Civil War.

Stitt and his cohorts planned meticulously, doing a good bit of reconnaissance and outfitting themselves like commandos, complete with headset radios and night-vision goggles they bought from a commercial outfitter. They used a water cannon to shoot a line over the flat top of the monument, then used the line to pull up a climbing rope. Stitt then used a system of knots to scale the vertical climbing rope.

"I got halfway up, then had to come down when the MPs [Military Police] came around. Then I climbed the whole thing."

After tying himself in at the top, Stitt unrolled a long vertical banner with "GO ARMY" and an MIA symbol on it. The idea was that the sign would be visible to most of the corps when the cadets came outside for breakfast formation at 6:30. But the Military Police, on their regular patrols, spotted Stitt's ground crew. Those cadets, perhaps thinking of the adage "Live to Fight Another Day" scattered, leaving Stitt alone on a tiny footing, looking for the first signs of dawn from his perch. Although the MPs hadn't spotted him, he couldn't get down without help. After half an hour, he called down to the startled patrol. The worst part was that they made him bring the banner down with him.

"No one got to see it or even take a picture of it," he says, disappointed.

It is time for lunch, and the new cadets have been issued their first MREs (Meal, Ready to Eat), the bagged, freeze-dried field rations. As with many of the things they do that are new to them, the new cadets defer to those with prior service. In this group where experience is measured in weeks, a new cadet with a year in uniform is an old veteran.

New Cadets Jay Koolovitz and Deborah Welle coach the third and fourth squads as they navigate the tricky waters of eating Army rations. Koolovitz was in radio repair before attending the USMA Prep School. (Most enlisted soldiers admitted to West Point spend a year at the prep school, preparing for college-level work.) He has a narrow face and bright eyes and is older than most of the second class cadets. New Cadet Deborah Welle was in Advanced Training at Fort Sam Houston, preparing to be a lab tech in the medical field, when she was accepted directly into West Point. She skipped prep school because of her academic record and high test scores. Welle, who smiles all the time, is the first person in her family to pursue a bachelor's degree. She is already thinking about her commissioning and graduation day: She wants her grandfather, a veteran of World War II Battle of the Bugle, to pin on one second lieutenant bar while her drill sergeant from Army basic training pins on the other.

In the brown packages, the meals look the same. New Cadet Joshua Renicker, a quiet, red-haired football player, studies the labels intently and eats everything he can peel out of the plastic. He is worried about losing too much weight during Beast, a common concern of the coaching staffs as well. Other new cadets shun some of the side dishes beans and peanut butter and hard crackers—as dry as road dust. Everyone eats the candy.

New Cadet Zachary Lange, a recruited hurdler from Minnesota, inspects his first MRE. He rips the top off a foil packet and squeezes the pasty contents into his mouth.

"What's that?" someone asks.

"I don't know," Lange says, chewing thoughtfully.

The upperclass cadets sit in a group nearby and talk about the most memorable event of Beast so far. They agree that yesterday's was a toss-up. At a swimming test, one new cadet, male, stripped down completely to put on his bathing suit. In a mixed group. Another new cadet took off his shirt to reveal a decidedly unmilitary nipple ring. (Body ornaments are supposed to be removed.)

As they finish their meals and fold the plastic wrappers, the new cadets talk about who is resigning. ("Quitting," a word heavy with judgment, is the most commonly used term to describe leaving the academy. No one ever says, "so-and-so wants to transfer to the University of Oklahoma.")

Rob Olson attributes these early resignations to the schedule of the first week of CBT: The new cadets spent most of their time standing in line at issue points to draw equipment or be fitted for uniforms; attending welcome lectures by the Superintendent, the Commandant, the Dean; taking tests for advanced-placement courses. They spent the July 4th holiday on guided tours of West Point's historical sites; they go to chapel on Sunday. There was none of the adventure promised in the recruiting literature. There was no time off, just busy work and constant commands in a strange new jargon.

Olson thinks it's critical to get the new cadets into the field and let them try something new. "I tell [the cadre] that if you can sponsor two or three consecutive successes you won't lose one new cadet [to resignation]."

As the new cadets finish lunch, the talk moves quickly to the day's big test: the gas chamber. This small block building, which sits some thirty yards from the break area, will be flooded with CS, the code name for the military version of tear gas. The new cadets will go in and, once inside, remove the mask. The stated purpose of the exercise is to give the trainees confidence in their equipment, specifically, in the protective mask. But there is another purpose here as well.

West Point believes that development takes place outside a person's comfort zone. Going into a small building filled with tear gas is outside the experience of most of these young people, and it is definitely outside their comfort zone. It is something completely new, and thus it is one of the essential elements of the West Point experience: a chance for the new cadets to stretch themselves.

"There's a priest up there," one of the cadre members tells the new cadets as he points up the hill to the chamber. (One of the chaplains is visiting the training site.) "But not because we think you're gonna die or anything."

New Cadet Marat Daveltshin is from Kyrgyzstan, the first "allied cadet" from a former republic of the Soviet Union. Stocky, with blue eyes and a trace of blond hair left on his scalp, Daveltshin is twenty-one and already has a university degree. He told Major Olson, "My country set me up for success."

Daveltshin listens intently to every instruction in English. He came to America two days before R-Day; up until then his whole experience of speaking English was in school. As the new cadets wait in formation, Marat speaks Russian with Ben Steadman, a new cadet who was an Army linguist with a Russian specialty. Olson says Steadman was put in the squad so that Daveltshin would have someone to help him navigate the demands of Beast. Steadman tries out his language skills; Daveltshin helps him with pronunciation, then compliments him.

"He'll go back and be a general," Steadman says.

"Not right away," Daveltshin corrects him. "It will take five or six years."

Cadet Bob Friesema, the tall, earnest redhead from Wisconsin, admits to being nervous. As his classmates chatter and a cadre member briefs them, he glances around at the smallish building and fingers the strap on his mask carrier. Friesema is one of the quieter members of Grady Jett's squad, though because of his size (he is six four), he cannot help but be noticed. He listens intently to every word of instruction, and he gives the impression that he is afraid he will do something wrong.

The plan is this: The new cadets will enter the chamber from the back door and be asked a few questions—name, hometown, the mission of the Military Academy. The idea is to get them to breathe while they're inside, to see how well the mask filters out the gas. Then they'll be told to remove the mask and say something else. This is when the gas will start to sting the eyes, the back of the neck, the face. Sweat exacerbates things, and everyone is sweating. The gas will cling to clothing and hands, so the new cadets are warned not to rub their eyes. They are also told that some people are more susceptible than others; some will come out vomiting, others will blink a few times and move on. Standing in ranks, the new cadets look at one another, as if they can determine who will be most affected.

When it comes to something unpleasant, the only place for a leader is up front, so the chain of command goes in first. Platoon leader Dave Hazelton, a firstie, says he'll do ten push-ups (outside the chamber) for every new cadet who can recite the mission of the Military Academy after removing the mask. Hazelton, Platoon Sergeant Stitt, and the squad leaders go in. All but Stitt come out.

The new cadets go in the back door, then come out the front coughing and gagging, several of them trailing streams of vomit. But they do not touch their eyes with their hands. A medic stands by for any serious reactions; most of them just walk around quickly, letting the air push the CS off of them. One new cadet, clearly in awe, comes out talking about Platoon Sergeant Greg Stitt, who has stayed inside the chamber. He pulled off his mask and is doing jumping jacks while he recites various pieces of new cadet knowledge.

New Cadet Omar Bilal is from Capitol Heights, Maryland, near Washington, D.C. When asked by a classmate if the chamber was bad, he flashes a smile and says, "Nah, the pollution in D.C. is so bad this didn't even bother me."

Pete Haglin, who is determined to follow his father into the field artillery, spits and coughs. "I have so much acne medication on my face all the time I didn't even notice if this stuff burns; my face burns all the time."

Bob Friesema comes out yelling, as if he'd just scored the winning touchdown. When his squad leader, Grady Jett, asks him how it was, he answers, "That was great, sir!"

Chalk up one of the successes that Major Rob Olson and the chain of command are looking for.

At the end of the training day, the new cadets move off the hilltop on foot, following a narrow blacktop road that takes them down to a large clearing where they will eat. The evening meal is already there: a dozen insulated metal cans, large steel canisters, called silver bullets, that contain cold drinks. There are stacks of paper products, trays of bread, and water for hand-washing. The cadre files through first, washing their hands, then taking their places to serve the food. The new cadets follow, helmets on, chin-straps buttoned, rifles slung at diagonals across their backs.

It is a comfortable evening, not too hot, and the new cadets spread out in clusters on the grass. Grady Jett, the Army football player who emphasizes teamwork above all, says a few words to Steadman, his prior-service new cadet. (Like many of the experienced new cadets, Steadman will often act as an unofficial assistant squad leader.) Steadman collects his squadmates and they sit together.

Pete Haglin is buzzing with excitement over the day's training and already looking forward to mountaineering the next day. The anticipated highlights: a competition to see which squad gets the best time building a one-rope bridge, and a rappel down a seventy-five-foot cliff. Haglin talks happily around bites of food, but his enthusiasm is not infectious.

Jacque Messel, the only woman in the squad, sits next to him, cross-legged, the muzzle of her rifle resting on one knee, a loaded paper plate on the other. She doesn't look up as Haglin continues his patter, until the cadre on the serving line start yelling at the new cadets to hurry up and finish eating. Messel and the rest of her classmates have spent most of the last hour standing in some line or other: waiting to move down off the hill, waiting at the bottom of the hill while the cadre figured out which platoons would eat first, waiting in line to wash her hands, waiting in line for food. As soon as she got her food someone was yelling at her to hurry up and eat. There is plenty to eat, she says, and almost no time to eat it.

Haglin finds out that Messel's father, like his, is a West Point graduate.

"My dad was FA [Field Artillery]", Haglin reports. "When I was born he put away all the gold stuff [uniform insignia] for me so that I can have it when I'm commissioned."

Messel, unimpressed, responds by pushing a lock of hair behind her ear.

"I used his cadet saber to cut the cake at my high school graduation party," Haglin says happily between bites.

"My dad saved his saber for me, too," Messel says without enthusiasm. She rolls up her paper plate, which has plenty of food still on it, picks up her rifle and stands.

"This place is not for me," she says.

As they finish eating, the new cadets are herded up a small grassy hill above the clearing where they ate. The entrance to the mountaineering site, where they will train the next day, is nearby.

The squad leaders and platoon sergeants move their charges into a large rectangular formation on the hillside. Their challenge is to get the entire company to set up neat rows and files of pup tents on the grass. Thirteen hours into their day, the cadre are about to be tested.

The new cadets drop their rucksacks on the ground. Many of them also ground their weapons, helmets, canteens, and ammo pouches. Their camouflage shirts are sweat-stained, but they are relaxed and talkative, and soon the occasion turns social.

"Why can't we have a camp fire?" one of the new cadets asks a classmate.

"Because the enemy will see where you are."

Each person carries a "shelter half"—half a tent. Team up with a buddy, button the two halves together, snap together the tent poles, run the guy lines, pound in a few stakes, and you have a neat tent with a triangular cross section.

But as they unroll the tents, many new cadets look at the equipment as if seeing it for the first time. Others are not all that interested in getting things set up. Not all of the second class squad leaders, for that matter, seem engaged with the task. There are few instructions from the company leadership. Many of the cadre do not realize just how much guidance the new cadets need; others are just ready for a break, too. They sit on the grass and watch the dusk roll in.

Some of the squad leaders jump to the task. Grady Jett is one of those. He shows his new cadets how to lay out the tent, how to space the rows and line up the tent stakes so that everything will be neat and orderly. Other cadre members use the time to visit friends in other platoons. Another squad leader, visible in the failing light because of her startling blonde hair, is working on her tent when a male cadet appears and—though she doesn't ask him to—begins to help. His squad of new cadets is not close by.

Many of the cadre members have removed their helmets, substituting the more comfortable camouflage soft cap. The new cadets take their cues from the cadre and begin adjusting their uniforms for comfort.

Another squad leader stands at the end of the row of tents and watches her two prior-service new cadets show their classmates how to set up a tent. Other new cadets from her squad stand and walk by on their way to the latrine at the bottom of the hill. One is wearing a helmet, the other is not; the one with the helmet has his weapon, the other new cadet does not. They pass just a few feet in front of her as they walk happily down the hill, chatting amiably in the dusk, like kids at summer camp.

This is a hazard of using trainers who are themselves just barely removed from basic training. If this had been Army basic training, the cadre would have been a drill sergeant with ten or more years in the service. If this had been army basic, the trainees would have learned a pointed lesson about never, never, *never* leaving your weapon behind.

Darkness is falling, bleeding the color from the scene; soon everything is gray going to black. Olson and Turner, the Tacs, sit on a low concrete wall, a remnant of some long-gone storage shed. Olson calls Cadet Josh Gilliam, a junior and the harried First Sergeant of Alpha Company. "What's your most precious resource right now?" Olson asks in his patient way.

Gilliam pauses before answering. He is in a complete uniform, rifle slung over his shoulder, helmet on, chin-strap buttoned. This is as much because he hasn't stopped moving as because he wants to set an example. "Daylight," he answers.

"Good. And what's your most important mission right now?"

Beside Gilliam about half the company is gathered in clusters on the slope. There are lots of people, including cadre members, standing around doing nothing more than enjoying a break. The other half of the company is still at the bottom of the hill, finishing supper.

"My most important mission is to get everyone fed," Gilliam says.

"OK," Olson answers. He doesn't indicate whether or not he agrees.

"You've got to prioritize, allocate resources, backwards-plan," Olson says, ticking off on his fingers. He does not give Gilliam any more specific instructions than that; it's up to Gilliam to figure out what to do. This is a learning environment, not a combat situation.

"We're not going to Bosnia tomorrow," Olson says.

Olson's plan: It is more important for Gilliam and his cadet NCOs to learn their business than it is for the new cadets to have a perfectly laid-out bivouac. There are some things, however, that are not negotiable. Anyone involved in leader development must know where to draw that line. All the new cadets must eat, for one thing. And the cadre must know where every new cadet and all the company's equipment is. The most sensitive items are the weapons.

Ten minutes go by, and not much changes in the scene, nothing more gets done. Turner sighs, looks up at the sky; it is dark enough now so that facial features are beginning to disappear. Turner gets up and finds Gilliam, who is still running around. Turner points out that, in a very short while, rucksacks and tent poles and weapons will become invisible in the darkness, blending into the background of tall grass. Nothing will derail a training exercise faster than a lost weapon. Like safety issues, weapons accountability is a showstopper; a lost weapon can end the career of the commander or NCO who didn't take proper precautions.

This is an area in which the officers will make pointed suggestions, will give direct orders.

"Why don't you set up your own tent?" Turner says. "That way the platoon sergeants know where to find you. Tell them you want a report on weapons accountability; have them come to you."

"I'm not comfortable asking the platoon sergeants to come to me," Gilliam says.

This is another way West Point is not like the Army, another challenge to the Commandant's plan to have cadets act in the capacity of NCOs and officers. In the Army, the First Sergeant would be senior to the platoon sergeants and would not hesitate to have them jump for such an important issue. More than that, the platoon sergeants would be experienced enough to know that the first sergeant carried heavy responsibilities; they would do everything they could to help. But Gilliam and the platoon sergeants and squad leaders, all the cadet NCOs, are classmates, all of them second class cadets. Olson is fond of saying that leading peers is one of the toughest leadership challenges. And so it is. Gilliam's feet are being held to the fire, with this captain breathing down his neck, with darkness coming on, with new cadets and some cadre members wandering around the bivouac, some with weapons, some without.

Turner says later that he came closest to losing his temper at that moment, when Gilliam started balancing his own "comfort" with something as important as weapons accountability. But he didn't. Instead, he reminded Gilliam about the chain of command, about who works for whom and what the priorities are.

"If you're not sure of the technique to use, ask Sergeant Bingham or Sergeant Mercier," Turner says, naming the two regular Army NCOs who are on board to train the cadet sergeants.

"But get accountability," he adds firmly.

In a few minutes, Gilliam has his report: All the new cadets and all the weapons are accounted for. Olson and Turner, whose reputations and careers were most at risk as darkness fell and the weapons remained uncounted, had been remarkably patient as they let Gilliam figure out how to do his job.

At the bottom of the hill, Master Sergeant Don MacLean, the senior Regular Army NCO in cadet basic training, watches as the last cadets pack up the chow line. MacLean, who is put together like one of the Abrams tanks he has commanded, is concerned that CBT isn't difficult enough, that the cadre are too close to being new cadets themselves. The cadre can get too buddy-buddy in the interest of being an "inspirational leader."

"Cadets aren't used to thinking in terms of 'I outrank this person,' and so can't keep that distance," MacLean says.

This is not peculiar to West Point. Many soldiers experience this same conflict when they first pin on sergeant's stripes. *Yesterday I was one of the boys; today I'm in charge.*

A few days earlier MacLean came upon a cadet squad leader sitting on the floor with his new cadets, chatting as they shined shoes together. "I pulled him aside and said, 'They don't need your war stories right now. Tomorrow you're going to have to inspect, and if those shoes are screwed up—you're not going to be able to say anything about it.' "

A new cadet comes out of the nearby latrine, just visible in the fading light; he is not carrying his weapon. MacLean takes hold of the new cadet's suspender.

"Get your weapon and don't ever leave it behind again. Understand?"

The new cadet sputters his understanding. This is an epiphany. No yelling, no threats, no histrionics, just a big man stepping out of the darkness with a fixed idea that a soldier needs to keep his weapon at arm's reach.

Josh Gilliam's lecture from the Tac NCOs about weapons

accountability probably sounded a lot like what MacLean said to the new cadet he encountered. The Regular Army officers and NCOs assigned to Alpha Company know the score; the challenge is that the experienced leaders have to give the inexperienced leaders time to catch on. They have to say it, let it sink in, reinforce it when necessary, and stay out of the way.

Like a lot of Army training, the next day's mountaineering begins with a demonstration of how these skills might be used in combat.

The new cadets sit on huge boulders near the bottom of a sheer rock face, turning the open space into an amphitheater. Four soldiers—men from the 10th Mountain Division, a regular-Army unit that has been sent to help with this summer's training—sit on the rocks directly beneath the cliff. They wear their BDU shirts inside out and keep a sloppy watch on what's going on in front of them, thus playing the role of enemy soldiers who believe their rear is protected by the cliff.

The lieutenant in charge of the training site climbs high on a rock above the new cadets.

"To an untrained soldier a rock face such as the one you see before you would be an obstacle."

Four GIs appear at the top of the cliff, behind and above the enemy. Suspended from climbing ropes, they lower themselves head-first over the edge of the cliff. They hold their weapons in one free hand. When they are just above the "enemy," they open fire; in the confined space made by the cliff and the boulders, the blanks sound like explosions. Two of the guards "die" outright; the other two return fire. One of the attackers is hit. A few new cadets jump as he falls a foot or two, then stops and dangles head-down from the climbing rope, held only by gravity and by a loop of nylon line through the metal D-ring at his waist. After dispatching the remaining defenders, his buddies climb back up, unhook the wounded man, tie him to the back of one of the others, then rappel to the bottom. The young men scuttle about the rock face as easily as if it were flat ground.

On the deck, one soldier checks the enemy casualties, removing their weapons; another is on the radio, reporting the friendly casualty, the enemy dead, the successful completion of the mission.

"Hoo-ah!" the new cadets yell at the end of the little drama.

Before Alpha Company gets out on the high cliff, there are more basic skills to learn: how to tie the knots that will hold them up; how to fashion a "Swiss seat," the rope brace that wraps around the back and through the legs and from which their weight will hang.

Grady Jett's squad marches to a corral made up of a thick nylon rope stretched around some trees to make a forty-by-sixty-foot rectangle. Two cadet squad leaders demonstrate the knots; then, along with soldiers from the 10th Mountain Division, they move around inside the corral, checking, correcting, coaching.

Jett knows the importance of drills; his squad has been practicing their knots for two days. They each carry an eight-foot length of rope with them, and at every opportunity Jett puts them through their paces with the unfamiliar knots. When the cadet in charge of the training site asks, "Who knows what a square knot is for?," Jett immediately shoots up his hand. His new cadets follow instantly, proud that their squad is prepared.

Staff Sergeant Bielefeld, the NCO in charge of the training site, sits on a large rock some twenty feet above the new cadets. Bielefeld served with the army's elite Ranger Battalions before his stint with the 10th Mountain Division.

"West Point turns out good officers," he says, watching the low ground and the scrambling cadets. "So does ROTC. I only knew one dud West Pointer; he was a by-the-book guy."

Another NCO, who spent four summers training ROTC cadets at their Advanced Camp (military training) thinks that West Point cadets are at a disadvantage.

"They're not used to dealing with soldiers and NCOs; they're so isolated here. And they act like wild bucks when they get out. Of course, I've seen that with ROTC lieutenants, too."

His complaints are not unusual. Lieutenant Colonel Dave Brown,

'80, who commanded hundreds of brand-new infantry lieutenants at Fort Benning for their initial training, said the West Pointers had a more difficult time handling their newfound freedom than did the ROTC graduates.

"They're locked up for four years, then they get out; they have money, a car, freedom," another graduate said about new West Point lieutenants. "They go nuts."

The new cadets of Alpha Company first practice rappelling on a gradual incline, a sixty-or seventy-foot rock face at a forty-five-degree angle. The slope, and just about every other rock in the area, has been painted with the blue and red 10th Mountain Division shoulder patch, but the most common decoration is the black and gold army insignia that reads, "RANGER." The cliff faces, the bleachers, even the big rocks that sit alongside the trails, are all painted with oversize Ranger emblems.

The Ranger tab (a cloth patch some two and a half inches long by half an inch wide) is awarded to soldiers who complete a very demanding nine-week course in small-unit tactics. Particularly prized by infantrymen, it has a high place in the hierarchy of badges denoting toughness and skill. And, by regulation, it is available only to men.

Colonel Peter Stromberg, head of the Department of English, is a former infantry officer who earned his Ranger tab after graduating from West Point in 1959. Stromberg thinks the omnipresence of the Ranger tab at the training areas sends the wrong message to women cadets: Rangers are the ultimate soldier. Women can't be Rangers; therefore, women cannot be ranked among the best.

Lieutenant Colonel Kathy Snook, a professor in the math department and a member of West Point's first coed class, sees this as a leadership problem, something the Department of Military Instruction (which has overall responsibility for summer training) should "fix." "It goes along with a kind of mentality that says, 'If you're not infantry or armor or field artillery, you're not really in the army.'"

To her way of thinking, the people who talk about the army as if it included only the combat arms—specialties restricted to men—are denigrating the contribution of women.

There is an Orwellian flavor to the business of what a few words painted on rocks can mean. Ranger training was long known as the "Army's premier leadership school," because it taught a soldier to lead under exaggerated conditions of stress: hunger, fatigue, and continuous operations. It was nine weeks of "gut-check" leadership: no support groups, no consultants, just lots of stress and whatever courage a soldier can pull from deep inside. But according to the Public Affairs Office at Fort Benning, Georgia, where Ranger School is headquartered, the Army no longer refers to the training as a leadership course.

"We call it a 'small unit tactics course' [now]," a Public Affairs official says. "If you call it a leadership school, then it has to be open to anyone. You could hardly make a case for keeping women out of a leadership course, right?"

The painted signs on the rocks at West Point are thus a slippery slope that could lead to all sorts of arguments about the integration of women into the military. But none of the cadets appears to be thinking that as they try to master the art of rappelling down a cliff. At the top, the sergeants check the knots on the line before moving each cadet to the edge.

"Sound off, new cadet!" the sergeants urge.

"RANGER," the new cadets yell, the women just as loudly as the men.

Because this is risky training, everything is tightly controlled by the NCOs. No one moves unless he or she is responding to instructions. The new cadets are kept back from the edge, waiting in corralled lines, as if at an amusement park. This is called "positive control." The cadets become used to moving only when told, and when they are told to move, they do so instantly. Once conditioned that way, the idea goes, the new cadet standing on the edge of the cliff will continue to respond instantly, reflexively, even when the command takes the new cadet out into space.

The new cadets are directed to undo, then re-tie their Swiss seats, the rope girdle that will hold them up.

"C'mon, c'mon," a sergeant barks at them when they move too slowly. "We're not splicing DNA here."

There is a twenty-five-foot cliff and a seventy-five-foot cliff. The new cadets practice on the smaller one first. The sergeant in charge of the twenty-five-foot cliff points at the first new cadet in line and summons her forward.

New Cadet Deborah Welle, the wit of third squad, steps up and faces the instructor, who is a head shorter than she is. He is the picture of cool confidence and speaks in a firm, quiet tone, signaling that he is completely in charge of everything going on. He yanks on her Swiss seat to make sure it is tied correctly and is tight. If she falls, even if she lets go with both hands, she'll hang from this girdle of nylon rope.

Once she has hooked the climbing rope into the D-ring at her waist, she turns her back to the cliff and shuffles toward the edge.

"This is interesting," she says. There is a little tremor in her voice.

A moment later, the tremor is gone. Just as she'd been instructed, when she reaches the edge, Welle shouts, "On rappel, lane one!"

Down below, another new cadet takes the ends of the ropes in his hands. If she falls, he need only pull the ropes tight and she will stop falling.

"On belay, lane one!" her partner shouts back.

"Loosen your grip with your right hand, let the line play out, and get into that L shape," the NCO tells her.

The first move is the trickiest. She bends at the waist and leans out over space until her legs are perpendicular to the rock face, her upper body still upright. To do this, she must move her feet a couple of inches over the edge of the cliff.

"This is a first," Welle says almost under her breath. "No jokes."

And over she goes.

At the bottom, smiling, she just remembers to shout, "Lane one, off rappel!" as she clears the rope.

Over on the seventy-five-foot cliff, New Cadet George Elias, who has been Alpha Company's model new cadet, stumbles a bit on his first bound. When he swings, his feet are below him, stretched toward the ground instead of toward the rock. He crashes into the wall.

"Bigger bounds," an NCO calls to him from the top of the cliff. The men at the top lean daringly over the edge, hanging by the safety lines at their waists. "Bigger bounds."

Elias tries it again, gaining confidence. When he reaches the bottom, he smiles and announces, "That was great. Especially when I smacked my face into the wall."

Jacque Messel is more timid; she tries to keep her body close to the rock wall as she goes over the edge, a beginner's mistake. She winds up almost vertical, balanced on her toes, an unstable position. The NCO at the top leans over the edge, coaching her until she gets her legs in front of her, her hands braced on the line. She manages a couple of small bounds. On the ground, she backs up until the end of the rope slides through her D-ring, then smiles as she pulls off her thick leather gloves. She is sweaty and her uniform is covered with dust, but she no longer has the dark look she wore when she decreed, at last night's meal, "This place is not for me."

After mountaineering training, the new cadets move back to the barracks to learn another lesson about soldiering: For every hour a soldier spends in the field, he or she will spend two or three hours cleaning and repairing field equipment. Alpha Company gathers in Bradley Barracks, named for Omar Bradley, Class of 1915.

Jett directs his squad to one of the biggest rooms available, a three-man room belonging to New Cadets Ben Steadman, Pete Lisowski, and Tom Lamb. Within a few minutes of their gathering, the floor is covered with twenty pairs of assorted shoes and boots, all of them brand new and waiting to be spit-shined; ten rucksacks; fifty-some tent pegs; a pile of shelter halves; six cans of black shoe polish; eight or nine canteens; five pairs of running shoes; one bottle of nail polish remover; two cans of shaving cream (one empty); about two pounds of grass and dirt (brushed off the field equipment); and the ten harried members of Jett's squad.

They got to the barracks just ahead of a July thunderstorm that left the sky a steel blue. Because Jett, the Army football player, stresses teamwork above all, they have assembled in one humid box of a room to clean their equipment in assembly-line fashion. In doing so they're trying to win a small victory from the ever-present enemy: the clock. You can't actually hear it ticking (everything is digital), but they all know it's there.

Two new cadets hold shelter halves up against the wall while a third whisks off grass and dirt with a fingernail brush. Future artilleryman Pete Haglin stands at the sink, dirty tent stakes piled on his right, clean ones on his left. He moves the dirty ones under the tap and into the clean pile. Pete Lisowski sits inside one of the big wooden wardrobes that line one wall of this room; he uses a brush to scrape dirt off rucksacks and foam sleeping pads.

By 9:00 they're attacking the boots and shoes that cover the floor like flotsam. They trade theories about the best way to strip the finish off so they can spit-shine the leather. One works with nail polish remover, another smears shaving cream on the toes of his shoes, still another uses a stiff brush. Marat Daveltshin, the new cadet from Kyrgyzstan, dabs great globs of polish onto his boot. He has clearly decided success is a matter of thick coverage.

Jett hurries into the room, wearing the summer white-over-gray dress uniform. On his arm is a brassard that reads "Duty Driver." He carries a two-way radio in his hand. Tonight he is part of a detail that mans the command post known as Central Guard Room; in between driving runs he has come to check on his charges. As soon as he appears the new cadets begin firing questions at him.

"Sir, does shaving cream work better than nail polish remover for taking off this finish?"

"Sir, what about using a brush?"

"Sir, do you *have* to take off the finish first?"

"Is this right?" Bob Friesema asks, holding up a shoe and making little circles of polish with one finger stabbed inside a piece of T-shirt. Jett steps amid the debris, handing out advice, offering suggestions, reminding them to drink lots of water. The new cadets are going through what a social scientist would call "acculturation" as they master one of the more mindless tasks connected to military appearance. Jett, who is responsible for everything that happens here, even though he is on guard duty, is getting another lesson in leadership.

"You've got a TA50 layout day after tomorrow," Grady Jett tells his people as he pokes at the piles of field equipment. "Get your gear cleaned up and store it over your closet. You've also got shoe and boot inspection tomorrow."

Another change in the demeanor of Beast Barracks: Jett doesn't yell at the new cadets, doesn't kick their equipment around the room or berate them because they don't already know how to do the task or because they respond slowly. He is all about getting them to do the work.

His warnings prompt another flurry of activity as the new cadets check watches and gauge how much time they can devote to polishing now, and how long before reveille they'll get up tomorrow to continue. Jett doesn't need to yell; the unfamiliar task and the unforgiving clock provide plenty of stress.

Ben Steadman, the former Army linguist, reaches into a dresser drawer where brand-new T-shirts are stacked in neat piles; he pulls one out and rips it up to create more shine rags. Someone else, hurrying, knocks over the bottle of nail polish remover, and suddenly the room is filled with its strong smell. Jett tells them to remove their athletic shoes and air their feet out; in a moment they are all barefoot, toes blackened by new combat boots. Tom Lamb, a soft-spoken blond from Washington, has wicked blisters on the backs of his heels, silver-dollar-sized patches of raw flesh. Jett takes a look, then sends Lamb to see the medic. The new cadet returns in a few minutes with a tiny piece of moleskin to patch the wound.

There is little in the room that is personal. Steadman's bookshelf holds an annotated Bible and a text in Russian. There's a shiny new edition of *Bugle Notes,* the plebe bible. Finally, there is an Army-issue *Manual of Common Tasks*, the sourcebook for the soldier skills they must learn this summer. Lamb, who spent a year at the University of

Portland, has the issue books plus Kahlil Gibran's *The Prophet*. Each of the three new cadets who live in this room also has the allotted one photograph, maximum size, eight by ten. The big frames are filled with smaller pictures of friends and family. Two young women, their white bikinis stark against tanned skin, smile out of the frame on Lisowski's desk.

When the platoon sergeant, Greg Stitt, enters, the new cadets have another round of questions about spit-shining. They're the same questions they asked Jett, phrased in a slightly different way, as if to say, "I know there's an easy way to do this and you guys are just holding out on us." The new cadets address the cadre as "sir," but they aren't afraid of them; they're afraid of being unprepared for inspection.

Jett notices that two new cadets have left: Omar Bilal, a recruited football player from Maryland; and Marat Daveltshin, the allied cadet from Kyrgyzstan.

"Are you guys a team if two of you are off somewhere else?" he demands.

One of their number scurries to find the lost patrol; the two missing appear moments later, arms full of equipment and shoes, and crowd into the hot room.

"Bilal sat by himself at breakfast [in the field] this morning," Jett confides. There are several reasons Bilal might separate himself. He is the only recruited athlete in the squad, the only black, and he is a year older than most of his classmates. Jett isn't alarmed, but he is aware of the kinds of things that can splinter a team. His experience with Army football the previous season underscored this point.

"We didn't really come together as a team, and it showed in our performance."

Jett steps into the relative cool of wide hallway.

"Teamwork is the only way to make it through this place. The whole Army is a big team," he says, precisely echoing the Army's leadership manual. The team identity he's determined to create extends even to how the new cadets put on their uniforms.

"I make them get in a room together and check each other off. When we first started it was clear that they worried more about themselves. I want them to come out together; even if they know something is wrong, they're all wrong together. The first time I saw that they were working together was when one came out in the hallway with his shirttail untucked. I corrected him and dropped him [for push-ups], and the whole squad dropped."

The minute attention to uniforms and military courtesy is exaggerated in the barracks, where life is much more formal than in the field. It was in the barracks, in private and mostly out of sight of the officers, where the most vicious hazing used to take place. So it is here that the differences between plebe life then and now become most apparent.

Out in the hallway, new cadets walk briskly, but they do not "ping" (an exaggerated walk, like race-walking, that made plebes look like windup toys and led to shinsplints). When they look for a room number, they look around the hallway; twenty years earlier, such movements would have elicited screams of, *"Why are you gazing around my hallway? You want to buy this place, beanhead?"*

Of course it's easier to find things—like numbered rooms in a long hallway of identical doors—by looking around. But back then, it just wasn't allowed. Period.

Under the old system upperclass cadets frequently appointed themselves unofficial "gatekeepers." They even went so far as to mount campaigns to "run out" certain new cadets, singling out for extra hazing the ones who—in the opinions of these nineteen-and twenty-year-olds—didn't quite make the grade. They kept up the pressure until the new cadet quit. And if a plebe sometimes got run out because his voice was too high or he smiled too much or he didn't have what it takes—whatever that is—well, that was the price of doing business.

Now the cadre's job is to train the new cadets and make sure they perform the tasks to army standard. So Jett gives his squad instructions, then leaves them alone to do the work. Because he has to return to Central Guard Room, he has asked another cadre member, Alisha Bryan, to keep an eye on his squad.

Bryan is the Alpha Company counselor. Her job is to talk to new cadets about what's troubling them, to give them a sounding board, a place to vent without having to involve their squad leaders. Bryan seems older than her twenty years, with a calm, confident air that suits her job. She has wide, dark eyes and wears her short hair pulled back in a ponytail. Tonight she wears the yellow physical training shirt of the Beast cadre and shower shoes on her feet.

Kevin Bradley, Alpha Company's commander, thinks the counselors, who are not in the direct chain of command, are part of the problem.

"They're not much help," he says. "It's better than last year, when the counselors used to write reports on the cadre. Now they live with the company. But she can't tell us what goes on in counseling, because of confidentiality. The new cadets see her as someone they can talk to without having to call her 'ma'am.' Everyone calls her the 'milk and cookies lady.' "

Bradley's comments are not just a knee-jerk reaction to what some people see as an invasion of political correctness and sensitivity training.

"It should be the squad leader's job to counsel the new cadet," Bradley says. "We have to learn that. Besides, she's no more qualified to deal with new-cadet problems than are any other members of the chain of command."

More than that, Bradley sees some of his cadre becoming lazy. "Some of them are getting into the habit of telling the new cadet, 'Well, go talk to the counselor.' They think it lets them off the hook."

Inside the room where Jett's squad works, Jacque Messel sits with a brand-new pair of dress shoes on her lap. She checks her sports watch every few minutes, waiting for her 9:30 appointment to meet Bryan. Messel, whose father played such a big role in her entering Beast, has been thinking about leaving.

Bryan and Jett shake their heads at parents who push West Point on sons and daughters. As they stand in the hallway, Lange, the recruited hurdler from Minnesota, is in the Tac's office, on the phone with his father.

"He wanted to go to the University of Minnesota," Bryan says. "His father basically told him don't come home [from West Point]."

Major Rob Olson is also standing by in the hallway outside his own office in order to give Lange a little privacy. Olson says that Lange has an older brother in the Class of 2000; the two haven't spoken in a year.

"*That's* a sign of a healthy relationship," Olson says sarcastically.

Lange finishes his end of the conversation but doesn't hang up; his father wants to talk to his son's Tac. Olson gets on the phone, delivers a couple of polite "yes, sirs" and hangs up.

"The guy is a dentist," Olson said. "And he gets on the phone with me and says, 'This is what needs to happen out there, Major.' "

Olson seems mildly surprised that Dr. Lange felt qualified to tell him how to run his unit. "The guy's a prick," he says. But the Tac is more upset at the cost to New Cadet Lange, who is isolated and unhappy and now embarrassed.

Back outside Grady Jett's room, New Cadet Daveltshin, from Kyrgyzstan, passes Jett and Bryan, his face intent. He says something in a thick accent that sounds like "Engo don, tsir."

Jett explains that the squad motto is, "Ain't goin' down, sir," which is from a Garth Brooks song. Jett says Daveltshin had a hard time with the idiom.

"He was at attention against the wall and he kept leaning forward to catch everything I was saying." Jett demonstrates the pose—head cocked to the side, bent over at the waist—that Daveltshin used to try to grasp the strange words that probably weren't in any textbook he'd ever studied. "I told him, 'You're at *attention!*' "

Jett and Bryan laugh, but they admire the guts it takes to come to a foreign country and an alien culture to try something so difficult. The second class cadets talk about finding someone—a professor, an upperclass cadet—who speaks Russian so Daveltshin can talk to someone in his native tongue. When the new cadets were allowed their weekly ten-minute call, Jett loaned his own telephone calling card to help Daveltshin phone home half a world away.

A new cadet from another squad appears, his hand full of laundry tags. He squeezes himself between Jett and Bryan and the room full of his classmates. He is clearly lost. In the past the hallways were always a free-fire zone, and a second's hesitation or hint that you didn't know exactly where you were going or what you were doing invited all sorts of unpleasant attention. The new cadet, whose shaved head and oversize, government-issue glasses make him look even more helpless, stands at awkward attention.

"What are you doing?" Jett asks.

"Sir, I'm trying to deliver these laundry tags to the upperclass cadet rooms."

"But you don't know where my room is. You just thought you'd come down here because I was standing here."

Inexplicably, the new cadet breaks into a wide smile. It's the wrong response.

"Smirk off," Jett snaps. "Did I say something funny?"

This is the kind of response that, in the past, would have drawn a half dozen cadre members into the hallway for the fun and yelling. But that doesn't happen. Instead, Bryan turns to the new cadet; he's huge, six four or five, and towers over her.

"Tell me what your mission was when you left your room," she says, leaning comfortably against the wall, her arms folded over her chest.

The new cadet gives a reasonably clear statement of his mission. Then she asks, "What was your plan?"

He didn't have a plan, so she talks him through what he should have done before venturing into the hallway. The new cadet stands at rigid attention, a giant tree. Inside his enormous running shoes, his toes wiggle furiously.

This is the point at which, many old grads would argue, the new cadet should learn a hard lesson, with lots of screaming to reinforce the learning point: Plan ahead. The new approach at West Point is this: He'll learn the lesson if he processes what is happening, if he thinks about it. And whether or not he thinks about it has always been up to the individual.

For her part, Alisha Bryan is learning how to teach someone a lesson—one all lieutenants should master. And she *isn't* practicing techniques she'll have to unlearn as a junior officer.

Fifteen minutes later Clint Knox enters the crowded room with instructions from the platoon sergeant on how to send out laundry (this will be their first time). When he shares these with his squadmates, they respond with questions.

"Do we separate dark and light in those little mesh bags?"

"Do we fill out a tag for each little bag?"

"How many sets of BDUs should we send out and how many should we keep?" Tom Lamb asks. "How long does it take to get them back and how many uniforms do we need between now and then?"

They don't know enough about the upcoming training to predict what they will need. Knox screws up his face and swears. Greg Stitt, the platoon sergeant, gave him about 70 percent of the information he needs.

When Jett and Greg Stitt reappear, the new cadets ask about the training schedule, about how long it takes to get uniforms back, about where to drop off the bags of dirty laundry and where to recover the clean stuff when it returns. Jett and Stitt are surprised by the questions. They didn't think through just how much information the new cadets would need; they forgot they have to walk trainees through everything.

Stitt shakes his head; he'll have to go around to all the squads again because his first instructions weren't clear. He explains the process, then shuffles out of the room, one lesson closer to being a lieutenant.

At 5:30 the next morning the new cadets of Alpha Company gather for their first instruction in Close Quarters Combat, the modern incarnation of hand-to-hand combat. The Tac officers and NCOs stand twenty yards behind the formation. The company first sergeant, second class cadet Josh Gilliam, runs by, papers fluttering in his hand. Gilliam is still moving at top speed, just as he was two nights earlier as he tried to organize the bivouac and get used to the idea that the platoon sergeants had to report to him.

"First Sergeant," Major Olson calls. Gilliam skids to a halt.

"How many sets of BDUs do the new cadets have left now that we've sent out laundry?"

Gilliam does a fair imitation of a deer caught in headlights. He glances left and right in the middle of the open field, as if the answer might be lying on the dewy grass.

"Sir, I don't know."

Olson chuckles. He could have warned the cadre about this last night, of course. But Rob Olson doesn't want to be the First Sergeant; he wants to teach Gilliam to be the first sergeant. Besides, if he stepped in, he would have violated one of his rules: Let the leaders do their jobs. That means taking a chance that, now and then, things will get screwed up.

"That's not quite the answer I was looking for," Olson says gently.

"Sir, I'll find out," Gilliam says, then races off.

Twenty yards away, someone is finally yelling at the new cadets.

"Welcome to Close Quarters Combat!" the instructors bellow.

"Hoo-ah," the new cadets yell back.

As the first weeks of Beast slip past, the Hudson Valley becomes more inhospitable. By this mid-July morning near the end of the first detail, it is as hot and sticky as the Deep South. Although it is only 8:00 A.M. the air is thick with moisture and annoying clouds of insects as the sixteen squads of Alpha Company move to a training site for a day called Squad Competition. More than a dozen stations are spread around an open area the size of three football fields. At each the new cadets will have to do some sort of event that involves teamwork and athleticism: running, lifting, climbing, carrying, jumping. The cadre members keep score, and all the numbers are posted on a big tote board in the middle of the field where everyone can see them. These are not, strictly speaking, military skills. This is more of a very athletic version of the kind of team-building exercises some corporations put their employees through.

Major Rob Olson stands with Kevin Bradley, the Alpha Company commander, who will turn twenty-one tomorrow. Since Bradley and his cadre are on-duty, working with the new cadets twenty-four hours a day, seven days a week, there'll be no time for a celebration. This day, as with every other in CBT, training started at 5:30 and will end, for the new cadets, at 10 P.M.; the cadre will stay up longer.

"Kevin," Rob Olson calls; Bradley comes jogging over. He wears his BDU cap pulled low over his eyes. The insects swarm around his eyes, ears and mouth, but he tries to ignore them as he listens to Olson.

"What are you going to do for the squads that win the competition?" he asks.

Olson's accent is not Minnesota, but the almost-Southern drawl that can be heard among career NCOs and officers from any part of the country. A staff sergeant from New Jersey sounds a good deal like a lieutenant from California. Delivered this way, Olson's questions never sound like a challenge. Whether or not it's an affectation, the result is clear: Cadets open up to him.

"Sir, we're going to let the squad leader buy them pizza to eat in the barracks."

This is a big deal. The new cadets gulp their meals down in the mess hall—while sitting at attention—or while sitting on the ground in the field. A little relaxed junk-food orgy will be a real treat.

"That sounds great," Olson says. "How about the squads that bomb? What are you going to do for them?"

Bradley, who stands with his hands clasped in the small of his back, shifts his weight slightly, then finally swats at a bug near his ear. Another long second or two later, he says, "Sir?"

"Well, what's our objective out here?" Olson says in the same tone he might use to ask, "How's that ol' huntin' dog?"

"Uh . . . build squad cohesion, sir."

"Sure. You've got sixteen squads, and some of them aren't going to do well. Somebody has to come in last place. So how are you going to build cohesion in the squads that do poorly? You're not just going to write them off, are you?"

Bradley knows the answer to this question. "No, sir."

"Good," Olson says. "Go figure it out and tell me what you come up with."

Bradley moves to the shade of some trees near the middle of the field; he removes his canteen from his equipment and takes a drink.

"This is the hardest job I've ever had," he says as he watches a squad of new cadets run an obstacle course.

"Major Olson doesn't give us any answers. We have to figure it all out on our own." Bradley isn't complaining, just acknowledging that Olson is making him earn his pay.

A little while later Bradley trots back to Olson and briefs his plan. When he shoots off on some tangent, Olson doesn't correct him. Instead, he asks a few questions that steer Bradley back on track. Olson doesn't have just one answer in mind, something he wants Bradley to divine. He listens and makes a few comments on Bradley's plan.

Olson, who has commanded hundreds of men from Korea to the Persian Gulf, could come up with an answer, probably a better answer than the twenty-year-old cadet in front of him. But as Olson has consistently maintained, he isn't here to do the cadets' jobs for them; he's here to make sure they learn how to do what's expected of them.

"The only way to do that is for me to get out of their way," he says.

Not all of the Alpha Company cadre learn their lessons as well as Bradley.

At one site the juniors who are supposed to be in charge are disorganized. The cadet giving the opening briefing repeats himself two or three times. When they get out the score sheets they realize that they haven't brought any pens to record the scores, and they have to borrow one from a new cadet. The new cadets are always being harassed about attention to detail and meticulous preparation, so it isn't lost on them that the cadre screwed up.

Because Rob Olson lets his subordinate leaders lead, he takes a chance that they'll "drop the ball" now and then. Olson is responsible

for everything that happens or fails to happen with these new cadets all summer long. If one of Olson's superiors shows up and finds
things aren't to his satisfaction, Olson will hear about it. (Colonel
Joe Adamczyk, the Brigade Tactical Officer and Olson's boss, is
famous—infamous among cadets—for nit-picking the details. And
he is always on the prowl.)

But Olson has decided to live in that scary place between always
doing things the safe way and taking a calculated risk to develop leaders. He accepts a bit of uncertainty as the norm. Olson's questions, his
Socratic method of challenging these young leaders is the flip side of
what most people want in leadership training: checklists, foolproof
methods, universal truisms, easy answers.

Throughout the summer, whenever one of his cadets says, "Sir,
we've got a problem," Olson never says, "Do this and this only." He
says, "OK, what are we going to do about it?"

To his way of thinking, whenever he throws it back at them, he is
sending a couple of messages: "I think you're smart enough to figure
this out", "I trust you to do this right", and, "You're worth my time
and effort, good enough for me to bother getting you ready for bigger
things." Those messages, more than anything else, are at the root of
"inspirational leadership."

PREPARE FOR COMBAT

Less than two weeks later the new cadets of Alpha Company have moved from pseudo-athletic team-building exercises to a more serious business. It is early August, and the squads gather in groups at the bottom of a dusty hill at Lake Frederick, some thirteen miles from main post. These veterans of five weeks of Beast are here to learn ITT, for Individual Tactics and Techniques: Army-speak for how to move under fire without getting killed. On the slope before them they can see coiled concertina wire, ditches, and other obstacles amid trees and tall grass. There are worn places where others have gone ahead of them.

Specialist Fourth Class Stubblefield, a soldier from the 10th Mountain Division, is the instructor at this site. His face, covered with camouflage paint, is almost as dark as his uniform. There is a machine-gun simulator (powered by compressed gas) hammering away just a few yards from where he stands. Stubblefield has a powerful voice that he's run ragged by screaming to be heard above the din.

"Whenever you're on my course, I want to see you carrying your weapon at the ready position," he tells the new cadets. Although he's been in the Army just a couple of years, he speaks with the confidence and authority of a senior NCO. The new cadets, dressed in BDUs, load-bearing equipment, and helmets, their faces also painted dark green, listen intently. A couple respond with a low, "Hoo-ah."

"Let me see your weapon, high-speed," Stubblefield says to a new cadet.

He clears the weapon—every weapon is loaded until it's cleared—then closes the bolt and the dustcover.

"This is the ready position," he says, grasping the weapon firmly with both hands, one finger on the trigger.

"Why is this such a big deal? Why do we stress this?"

A new cadet speaks up, giving the obvious answer. "So we're ready."

"Right. Keep both hands on the weapon. This shows you're ready. This says to the enemy, 'Go ahead, give me a reason to waste you.'"

"This . . ."

He slouches, rests the butt of the rifle on his ammo pouch at his waist. "This says 'I don't care.' And when you're pulling guard duty in Bosnia or in the Sinai, this makes the enemy think, 'I can get in there and plant a bomb.'"

Stubblefield hands the weapon back to its owner. In the rear of the gaggle of new cadets, two women stand side by side. One of them holds her rifle slung on her shoulder; it is almost as long as she is tall. "You know," she says to the other woman, "we need to take this stuff seriously, because there's no such thing as a front line anymore."

All morning the new cadets have been practicing the skills they will use on this course. In the open space at the bottom of the hill they've learned how to crawl with a weapon, how to provide covering fire to one another, how to keep low as they move under fire. Now they face the day's big test: an uphill course of several hundred yards that will have them crawling, climbing over obstacles, shooting, moving as a team, and covering one another's movements. There are no real bullets, but the course is rigorous, and plenty of people are watching.

When Stubblefield announces that it's time to go, squad leader Shannon Stein jumps to the starting position, calling for her new cadets to line up behind her. Stein, a five-foot-four-inch, hundred-pound bundle of energy, replaced Grady Jett for the second detail of cadet basic training. (There are two complete sets of cadre for the summer, which maximizes the number of upperclass cadets getting leadership experience and brings in rested upperclass cadets halfway through the summer.) Stein has dark hair and eyes and, beneath the camouflage paint, is fair-skinned. She is also a star on the women's soccer team, a recruited athlete whose heart was set on the Naval Academy until she visited West Point. The sleeves of her blouse are rolled into tiny cuffs just above the stock of her rifle; the smallest size is still too long in the arms for her.

At the start signal, Stein leaps forward, diving into the dirt behind a couple of piled logs. "Cover me, I'm moving," she shouts to her partner, then presses the side of her face into the dust—not close to the dust or near the dust or just above the dust—but deep enough to move a small bow wave of dirt before her. Flat on her stomach, she pushes herself forward with one leg, her helmet burrowing a path. She grasps her rifle by the sling, keeping it out of the dirt by draping it over her arm. This is the low crawl. The new cadets in her squad watch intently as they wait their turn. A couple of them murmur the five-point checklist they learned this morning as Stein goes through it before every move: Check the weapon's safety, the dustcover (which protects the bolt), check to the left, right, front.

When Stein reaches the first covered position, she rolls over and takes up a firing position, propped on her elbows, rifle forward, covering her partner as he moves. A few yards up the hill, the machine-gun simulator pounds the air like a string of car crashes. On the lane in front of them, a smoke grenade pops; the thick cloud hangs on the hillside in the heat. Once Stein and the other squad leader have moved forward, the first new cadets launch themselves on the hill.

The pairs zigzag through the dirt, moving from covered position to covered position, running in a crouch. The goal is to remain exposed for no more than three seconds, which makes the two-hundred-yard course a long one.

The noise level rises as each pair of soldiers begins moving. The ones covering fire blanks, and soon the lane NCOs are throwing hand-grenade simulators, which go off like enormous firecrackers. The attackers scream to each other over the din of firing.

"Cover me, I'm moving!"

"Safety, dustcover, left, right, front! Let's go, let's go, let's go!"

Soon there are two squads spread out on the hill. Slowing down is not an option: There is relentless pressure from the rear as more people join the assault. They must keep some distance between them ("One grenade can get you all!"). There is some shade provided by a few scrubby trees and tall weeds, but mostly the course is dusty and hot. The new cadets approaching the top look as if they've been working in a flour mill; their faces are streaked with sweat and runny camouflage paint. The dust turns black around their mouths, and they breathe it in with great gulps of air.

Up the hill, Stein dives into a hole, then crawls into the maw of a dark concrete pipe, like a sewer drain; her partner covers her over the top. She is small enough to squeeze through the space, but scrambling on the concrete bites at her knees. Her helmet and equipment bang against the sides as she works her way through fifteen to twenty feet of tunnel. When she emerges in the sunlight, her helmet has slipped down over her eyes. She pushes it back and raises her weapon to the ready position.

When it is her turn to move again, she approaches a field of tanglefoot: criss-crossed barbed wire strung two feet off the ground. Stein flops onto her back, lays her rifle across her chest and churns her legs to push herself under the wire. Here the earth has been ground to a fine powder that rises in clouds over her shoulders. She blinks away the sweat and the dust and powers through with surprising speed.

After ten or twelve minutes of tremendous exertion, the lead cadets are within a few yards of the "enemy" position, an eighty-foot trench near the top of the hill. It is deeper at one end than the other; the deep end is filled with green water. Stein and her partner plunge in, clear left and right, firing blanks at the plastic soldiers that occupy the trench. Once her buddy boosts her out, Stein checks in front of the trench for more enemy. Finished, she shoulders her weapon and turns back to watch her new cadets. She is muddy and soaked from the waist down, covered with white dust from the waist up; her face is a war mask of green camouflage and powder. Her breath comes in sharp spikes, but she manages to call encouragement to her charges.

"Let's go! Let's go! Let's go," she shouts, her voice high and surprisingly strong.

Allied cadet Marat Daveltshin thrashes beneath the tanglefoot, losing his way and getting hung up in the steel web. A soldier/instructor working the lane calmly coaches him.

"Open your eyes, Daveltshin!" Stein yells. "Look where you're going!"

When Daveltshin reaches the trench, Stein yells at him again: "Kill Ivan!"

Ivan is the generic name for a Soviet soldier. The trench is manned by little plastic silhouettes of "enemy" soldiers left over from the Cold War years. Each three-foot-high figure wears a red star on the front of his helmet. Daveltshin, who might have wound up an "Ivan" if not for the collapse of the U.S.S.R. butt-strokes the enemy with his weapon as enthusiastically as anyone else.

Bob Friesema plunges into the trench at the deepest end, helps his partner out, then tries to pull himself clear. The top of the trench is a rounded pile of dirt; he jumps up, but there are no hand-holds and he slides back slowly in spite of his exertions. He jumps again, slides down again into the rank water. He should be exhausted, but he becomes more determined, jumping higher still.

"C'mon Friesema," Stein yells. "You're seven feet tall. If I can get out of that ditch, you can."

Friesema's partner, out of the trench and on top of the berm, reaches back with one arm. Friesema grabs the offered hand and scrambles clear. He lays in the dirt, breathing like a beached whale, but remembers to scan his front, his weapon ready to meet more enemy.

Clearing the trench was supposed to be the high point of the exercise, the objective of the assault on the hill. Some of the new cadets become so engrossed in the idea of diving into the brackish water—and thus showing how "Hoo-ah" they are—that they forget to clear the trench; some pay no attention to the little plastic enemy. There are almost as many observers as there are new cadets in the trench—the lieutenant and NCOs from the 10th Mountain Division, a couple of Army medics, a dozen cadet cadre. The gallery of specta-tors defeats the effort to make the training realistic. Instead, the trench-clearing begins to resemble some fraternity initiation rite: Dive into the green water, run around, and shout.

When the last of her new cadets is through the course, Stein leads her squad off the hill and onto a paved road. The new cadets talk excitedly, trading war stories about how hard they ran or how deep the water was where they crossed the trench or how quickly they got up the hill.

At the bottom of the hill, they break off to fill canteens, check skinned knees and elbows as Stein watches them. "I've got this whole mother-father syndrome with them, you know?" she says, referring to her charges. "I'd do anything for them, but I'm hard on them, too, as they can tell by the number of push-ups and flutter kicks they do." She smiles. "They love those flutter kicks."

Although her days are long (she got up at 4:30, a half hour before the new cadets), she says being a squad leader "is definitely the best job."

"We're with them constantly, from reveille to taps without letup. The new cadets are constantly asking questions, they constantly need corrections, they constantly need watching out for. 'Have you changed your socks? Do you have water in your canteen? Button your chin-strap, fix this, fix that.' "

Part of her concept of leadership was formed this summer when she had an unusual—for her—experience: She failed one of the phases of Airborne School.

Stein went to Fort Benning, Georgia with a large contingent of cadets for parachutist training. Airborne School is divided into three one-week sessions: ground week, in which the trainees do a lot of conditioning and practice landings; tower week, during which they practice exiting an aircraft from a thirty-four-foot tower, then practice landing after dropping (in a parachute) from a two-hundred-foot tower; and jump week.

Shannon Stein, recruited soccer star and self-described "PT stud," washed out during the first week. She couldn't perform a "PLF," the parachute-landing fall jumpers do to lessen the impact of hitting the ground. A PLF takes a minimum amount of control and athleticism, and a jock like Stein shouldn't have had a problem with it.

"I was just used to falling on the ground in soccer, I couldn't learn to do it gracefully."

Stein was disappointed in the reaction of her "blackhat," her NCO trainer. (Cadre at airborne school wear black baseball caps.) "He just got disgusted with me and at one point said 'I don't even want to train you anymore' and he left me. I would *never* do that to one of my people. Later I found him playing cards with the other blackhats. He could have been training me. But it was a good experience for me. I'd never failed at anything like that before, you know? And I told this story to my squad: Here's this little athletic girl who goes down to Fort Benning and gets recycled through Airborne School—which isn't even really that hard. I could have gone home and had three weeks of leave and come back without my wings, or I could stay an extra week [to go through ground training a second time]. I stayed. I feel like I really earned my airborne wings. It's not like I got them out of a cereal box or anything."

"I told my new cadets about this because they were real nervous about going to BRM [Basic Rifle Marksmanship], but I told them 'You can get through this. If you put your mind to it, you can get through anything.'"

Stein tried the message out on Jacque Messel, who didn't qualify with her weapon the first time through.

"I told her, 'You can do it.' And she did. She was surprised, but I wasn't."

Messel, who is standing nearby, is disappointed that she has missed the morning's training because she had to see a doctor. If she misses too much, Messel will have to repeat the training the following spring while her classmates enjoy spring leave. But she is obviously sick, with a pale and exhausted look, and has been throwing up for a couple of days.

According to Olson, very few new cadets will try to get out of training by feigning or exaggerating sickness. Instead, they'll train when they're hurt and make things worse. Zachary Lange, the Minnesota hurdler, spent the morning of the assault course hobbling around with an infected ingrown toenail.

"If I walk on my foot just a little bit sideways," he says, demonstrating the angle with his hand, "I can make it [through the assault course]."

Lange is not alone. Stein's squad has a collection of bruises and scrapes and cuts. The medic patches them up and they head back to the squad. They are afraid to be left behind, but not because of the threat of a lost spring leave. They have become a team.

The sun is low and the sky a washed-out blue as the rest of Alpha Company comes off the assault course. They are muddy, soaked with sweat and the scummy water of the trench; the moisture makes clay of the dust that had been clinging to them. At the bottom of the hill they get a resupply of blank ammunition. Stein reminds her squad to drink lots of water. As she speaks, she taps a loaded magazine on her helmet, a trick to make sure the rounds are properly seated in the aluminum magazine.

Pete Haglin, who can't get enough, asks, "Ma'am, when do we get to do the cool laser-tag stuff?"

"That's next," Stein says.

The new cadets will learn how to wear and operate MILES

equipment, a set of sensors and low energy lasers that simulate firing and being hit by fire. The system is an electronic way of keeping score in a simulated battle. (The acronym means Multiple Integrated Laser Engagement System.)

At dusk the new cadets move off to another hill, this one an infiltration course in which they will "attack" another squad. For those who stay in the Army, this is only the first of many night exercises. With practice, trained soldiers can move and drive and even fly in darkness almost as easily as in daylight, but that kind of sophistication is years away for West Point's Class of 2002. For them, as for any soldier in the field for the first time at night, most of what goes on seems confusing and aimless. They wear themselves out thrashing around in circles through the thick underbrush until almost midnight. By the time they stop, they are soaked with perspiration; then the temperature drops as they stretch out on the ground to catch a few hours of sleep before the next day's training.

They are practicing the basic skills of the foot soldier. Even though most of the men and all of the women will wind up in some other branch of the Army, they will all have at least a limited understanding of what the infantry does, and how difficult it can be. By the second morning, they are beginning to show signs of just how far the lesson is sinking in.

After a hot breakfast, trucked from the Cadet Mess at West Point and served out of insulated cans, Stein's squad gathers in a sunny clearing for their first class of the day. They are all dirty; yesterday's mud and sweat has congealed into black and brown streaks that decorate hair, faces, hands, and clothing. A night of sleeping on the ground has done nothing to help them look refreshed.

The instructor is Sergeant Brust of the 10th Mountain Division. Brust wears a combat patch—he's a Gulf War veteran—and is not nearly as dirty as the cadets, though he has been on-site for days. He is not a big man, but he speaks with a calm authority as he stands before a portable easel. There are a couple of drawings illustrating how a squad of ten men moves under different conditions: when expecting enemy contact; when contact is possible; when contact is not likely. Brust uses a lot of jargon, some of which the new cadets may have heard before, some they haven't heard.

"Army doctrine is that we need a 3:1 ratio in the attack," Brust says. The new cadets stare blankly.

As the class goes on the sun climbs, and soon it is uncomfortably hot and close; the humidity clings to trees, grass, skin. The NCO talks about dead space, about masking fires. Every once in a while he'll explain one of the new terms, and there are others the new cadets can get from context. They sit in the sunlight, struggling to stay awake.

The eleven-man squad in the diagram is divided into two five-man wedges, with a squad leader in between. At the head of each wedge, the little circle is marked "TL," for team leader.

"You must stay twenty meters apart here," Brust says, touching the little black circles that indicate soldiers. "Here's the team leader. He trains [his soldiers], gets them ammunition, checks their feet, and makes sure they have dry socks, makes sure things are OK at home."

The new cadets blink slowly; none of this looks very difficult yet. You stand in a wedge. There is another squad at the site besides Stein's. When Sergeant Brust tells them to get up and practice the formations he's just talked about, they respond slowly. Suddenly Alpha Company's cadet First Sergeant steps out from where he's been watching and snarls at the new cadets to move quickly when an NCO gives them an order. There is a flurry of camouflaged arms and legs as the new cadets respond.

"That will not happen with us, do you understand?" Stein yells at her squad. She is embarrassed that the new cadets didn't show Brust more respect, and she's determined her squad won't do the same. "When he says move, you *move*. How you act out here is a direct reflection on how much pride you have in yourself."

When Stein talks to her new cadets, she has only one tone—harsh—and only one volume—loud. The approach is losing its effect. The new cadets respond with an unenthusiastic, "Hoo-ah."

Insects buzz. It's hot now, and the new cadets drink water in hopes of staying awake. Brust unrolls a chart titled "Prepare for Combat." There is nothing philosophical about it; the chart shows a list of equipment an infantry squad might carry. He reads it to them.

"Your grenadier carries forty rounds of 40-millimeter H-E-D-P," he says.

No one is taking notes. No one asks what H-E-D-P is.

"Your M60 machine gun can lay down nine hundred to a thousand rounds a minute. It's the most important weapon in the platoon."

"V-S Seventeen panels," he says, touching the chart. "Combat lifesaver bags."

The cadets follow the motion of his hand, as if he is a conductor.

"In your pre-combat inspection, you check for stuff that's going to make noise when you move, the water in your canteen, jangling equipment, that kind of stuff. You check their boots."

The new cadets doze in the heat.

"One of you might be my lieutenant one day," Brust says. "I might be your platoon sergeant."

Brust gets the squad up and into a wedge formation. Once he has them moving again, Brust becomes animated. Pointing to the woods ahead of them, he says, "OK, now we start taking fire."

The new cadets slowly go to ground and take up firing positions. They look out from under the brims of their helmets. Some of them lower their heads to the ground, as if sniffing the dust.

"Once we get fire superiority, we got to put some lead on 'em, then we can move."

The front team simulates firing at the enemy position; the five-man team in the rear, responding to the team leader's arm signals, starts to move to the enemy's flank. Two new cadets cross in front of their own men.

"No, no, no!" Brust shouts. "You're gonna get killed by your own guys if you step in front of them."

He talks for a moment about fratricide, about how easy it is to get killed by what the Army calls "friendly fire." In the Gulf War, friendly fire accounted for a whopping 26 percent of the 146 battle deaths.

Brust, in the age-old tradition of the NCO trainer, is trying to bring that lesson home to the next generation. But he doesn't tell them about the charred tanks, about the bodies burned beyond recognition, about the boys incinerated by high explosives. Instead, he gives them a five-minute break while he goes off in the woods to smoke a cigarette. Someone has told him he is not allowed to smoke in front of the new cadets.

If any of these young people, most just a few weeks out of high school, are startled to find themselves carrying automatic rifles and sitting though a class entitled "Prepare for Combat," it doesn't show.

"I'm sore everywhere my bones stick out," Barry DeGrazio says, rubbing his knees with open palms. All of the new cadets got beat up on the previous day's assault course. Everyone is bruised and a little battered. This is why they call the infantry a "bloody knees business."

"It's like football," Omar Bilal, the football player, offers. "The harder you do it, the less likely you are to get hurt."

Clint Knox, the dark-eyed graduate of a military high school, muses that he turned down an ROTC scholarship at Tulane—and gave up his summer—to sit out here with Sergeant Brust.

"And *Playboy* rated Tulane one of the top two schools when it comes to good-looking women." He says this seriously, as if quoting the *New England Journal of Medicine*.

Tom Lamb, who attended the University of Portland for a year, is happy to report that he had his fun. "I got it out of my system," he says, smiling at some memory.

When the talk turns to the first-aid training, Lamb surveys his squadmates, their faces thick with yesterday's grime and another coat of green camouflage paint.

"If one of you guys was dying, I'd give you mouth-to-mouth to save you." He pauses. "But you'd have to be dying."

Lamb has the gentle demeanor of a scholar, in spite of his GI glasses and trench-knife haircut. He was in Army ROTC at Portland; he is twenty years old.

"When you're twenty, you're not a teenager anymore," Bilal says. For these young people, twenty is old.

"Imagine Shakespeare," Bilal continues. His squadmates know he's referring to a classmate, not the Bard. New Cadet William Shakespeare, USMA '02, was an enlisted soldier in the Army and is several years older than his classmates, older than most of the cadre. "He's getting yelled at by people younger than him." Bilal shakes his head at the ignominy.

Barry DeGrazio talks about being in the fastest running group. New cadets are divided into black, gold, gray, and green running groups, based on their performance in an early physical fitness test, the first week of CBT. DeGrazio's group runs a sub-six-minute-per-mile pace up the steep hill behind the football stadium. It is, everyone agrees, an insane standard.

Pete Lisowski says, "I'm proud to be in the slow group."

They are all looking forward to school—and the end of Beast—and they are all nervous about college-level work. Clint Knox is concerned that high school was too easy and didn't really prepare him. He asks if cadets can be commissioned in the Finance Corps and says that his ambition is to go to business school.

This is the kind of talk that makes some old grads howl. They say West Point is about preparing leaders for the Army; it is not a place to polish a resume for business school. The problem with that thinking is that "West Point" looks great on graduate school applications, and every candidate knows it.

The enthusiastic Pete Haglin says that he got into West Point because he had good SAT scores. "But my grades weren't that great because I didn't do any work."

Haglin, the new cadet who received his acceptance to West Point after already making a deposit in housing at another school, is concerned about how he'll handle college work in the upcoming school year.

Shannon Stein comes back from the latrine and sits down with the new cadets.

"All right," she says. "Let's not waste time here." She begins quizzing them on their required knowledge.

"Hey, you," she says. "Tell me about the SALUTE [spot] report."

❖   ❖   ❖

Later, when they break for lunch, Stein sits with the new cadets again. She gives away most of her MRE, but scrounges for candy.

"I need a pick-me-up," she says. One of the new cadets tosses her the small packet of M&M's from his meal. Around us, other squad leaders have left their squads to eat alone; some upperclass cadets sit in a group.

"What do you think of me eating with you guys?" Stein asks. She is only two years older than her new cadets. When she looks for some affirmation that she is doing things right, she looks to the new cadets instead of to other cadre members.

A couple of them respond, "Hoo-ah, ma'am," which might be an endorsement, or might be a way to avoid the question.

"We're definitely eating Schade's deep-dish [pizza] when we get back," Stein promises them as she munches the candy.

"What kind of pizza did Cadet Jett get you?" she asks, looking for another yardstick against which she can measure herself.

Pete Haglin asks Stein, "Ma'am, why was last year's Beast so easy?" He already sounds like an old grad, and he hasn't completed basic training.

Stein doesn't question his assumption. "It goes in cycles, I guess. Only six people quit Beast last year."

"We had fifty people quit just during first detail," Haglin asserts.

In fact, Alpha Company has not lost a single new cadet.

"We had a kid who quit during the R-Day parade!" Stein says. "He saw his family [in the bleachers] and he just walked right off the field before we took the oath."

The new cadets are amazed, maybe even a little jealous. Now Stein is warmed up and has an audience. "Another kid in my Beast," she says, "had an uncle who lived around here somewhere. One day he took off in his running gear and ran to his uncle's house. It was, like, seventeen miles. You can tell how easy Beast was last year because those plebes were undisciplined."

Like most people their age, cadets are capable of dazzling generalizations, and the assumption underlying many of them is: My plebe year was the last hard one.

Just two years earlier Shannon Stein was herself a new cadet. The homesickness in her letters would not surprise anyone who has been around young soldiers, with their sudden appreciation of home life, their surprise at what they can accomplish.

"Let me tell you how great you have been," she wrote to her parents in 1996. "Every day I go to the mailbox, I have something. My buddies sometimes do not have a single letter."

After thanking her parents for some packages, she ends with lists of needed supplies, written in big, block letters.

"MOM, IF I DON'T TALK TO YOU . . . PLEASE SEND THE FOLLOWING: SMELLY STUFF, SPORTS BRAS 10 MORE, FOOD!, SHAMPOO/CONDITIONER."

She also wrote about the upperclass cadre: who was a good role model, who was not. Often the stories were couched in terms of who liked her.

"My squad leader is really nice to me," she wrote. The platoon sergeant: "He is so cool and he really likes me." During second detail, her squad leader was a prior service cadet, "one of those hotshot guys."

"My squad leader loves me," she wrote. "In fact, everyone in this detail loves me."

It is not surprising to find her using this language to describe her relationship with her leaders ("he likes me"). She was, after all, new to the business of senior-subordinate behavior. But Stein still seems confused during CBT '98. While she was no pushover for her squad, she may have been exactly what Master Sergeant Don MacLean had in mind when he talked about new leaders being confused about their roles, about not being used to thinking in terms of "I outrank this person."

She sees herself as a champion for her new cadets.

"My plebes don't do table duties," she says.

New cadets and plebes are required to perform weird rituals at the table. One new cadet is the "gunner," another the "hot beverage corporal," and another the "cold beverage corporal." Once everyone has been served, the gunner announces, in parade-ground tones, how many servings of food are left.

"I asked them," Stein says of her new cadets, "What is the purpose of a Mess Hall?' And they said, 'To feed people, ma'am.' 'Right,' I said. 'So eat.' "

She holds an imaginary plate over her shoulder and recites a litany that hasn't changed in at least twenty years.

"Sir, there are five servings of mashed potatoes left on the table. Would anyone care for more mashed potatoes, sir?"

She lowers her hands. "That's stupid. I don't make them sit at attention at meals, either. I think meals are for eating and should be relaxing. In my Beast I could never fully digest my food."

This makes her popular with the new cadets, but not with the other cadre members. "The platoon leader came over to our table one day when we had [an extra space]. I told him 'You can sit here, but they don't do table duties.' He's like, 'What?' I said, 'They don't do table duties and you can't haze them.' He left."

Stein's new cadets are happy to eat in peace, but they are already concerned about falling behind their classmates. There are dozens of table duties to be mastered, and the summer is the time to learn them well. Stein's new cadets are not learning these rituals, and they're worried about the academic year, when the table will be suddenly full of upperclass cadets scrutinizing every move the plebes make.

Ben Steadman has finished eating his MRE; he gets up, asks Stein where the Porta-John is.

"In the woods," Stein tells him. "But I don't know where you are."

Steadman walks off from the group without his helmet or his rifle. Two cadre members—one is the platoon sergeant with whom Stein has clashed—sitting close by immediately stop him, drop him for push-ups, and send him back for his equipment. They say nothing to Stein, and she says nothing to Steadman or to her classmates.

The talk drifts to the first-detail cadre. All of the new cadets admired Greg Stitt, the second class platoon sergeant who found his inspiration to come to West Point in the 82nd Airborne Division Museum.

"We learned from him that you take care of your people. He taught us something every day; he taught us how to get the job done," one of the new cadets offers. "This platoon sergeant is 180 degrees out,"

This draws no comment from Stein. In her presence, the new cadets refer to the second detail's redheaded platoon sergeant as "Lucky Charms," after the leprechaun on the cereal box.

When Stein gets up to go to the latrine, they talk about the time the second-detail platoon sergeant took over as their table commandant when Stein was absent for a meal.

"He said we were pronouncing Daveltshin's name wrong and he argued with us and wouldn't let us eat."

"He ate, though," Bob Friesema adds bitterly.

This is a major sin. Leaders do not eat before subordinates. The new cadets learned this first detail.

"He kept us in the hallway practicing facing movements for forty-five minutes," Tom Lamb says.

"You can learn a lot from bad examples of leadership, too," Friesema adds.

Lisowski says simply, "He sucks."

At the bottom of the hill where Stein's squad breaks for lunch, three captains wait for the training to begin again. The key to the Army's success, these officers agree, is to push decision-making down to the junior leaders. Captain Dave Grasso, a Green Beret who will spend the coming year preparing to become a Tac, tells a story about a corporal who was in charge of a team of four or five soldiers at a checkpoint in "BH" (Bosnia and Herzegovina).

"This crowd gathered and started throwing bricks at them [the GIs]. He went through his graduated response, on his way to using deadly force. But he knew that if he shot anyone it would be an international incident, so he kept his cool, even though the rules of engagement allow you to shoot if you feel threatened."

Captains Andy Groeger, who is also in the TOEP program, and Steve Patin, agree. Peacekeeping missions, the Army's stock-in-trade at the beginning of the new century, present a range of challenges.

"We have E4s and E5s [corporals and sergeants] manning road-blocks in Bosnia," Grasso continues. "We don't have enough captains and lieutenants and platoon sergeants to watch everything. Any area is a potential hot spot. They're making life-and-death decisions they might have referred to the chain of command if they had time."

A hundred yards away, beneath the hammering machine gun and the periodic *Blam!* of artillery simulators, another new cadet company is being introduced to the assault course and the gospel of infantry combat: if you can be seen, you can be hit; if you can be hit, you can be killed. There is very little shade, and the new cadets sweat under their helmets and heavy camouflage.

After the heat and the dust of the assault course, the cool dark of the bar in West Point's Hotel Thayer is inviting. The room is nothing fancy: cheap synthetic carpeting, bus-station furniture, lots of wood-grained plastic. At least the hotel sits on a bluff above the Hudson; its big windows keep watch over the channel and the low mountains on the eastern bank.

It is late in the afternoon; one of the few patrons is a tall man with a thick shock of white hair, electric blue eyes, and a plastic name-tag that reads "Jack Norton '41." Norton is at West Point for a meeting of the alumni association. He orders a Manhattan.

Norton, eighty years old the previous spring, still looks and talks like a soldier; he sits straight in his chair, long fingers curled around the stem of his glass. His West Point class ring—worn on the left ring finger, as it is by most graduates—is almost smooth with age. He is a remarkable storyteller, recalling details of sight and sound over more than half a century. In a few short minutes Norton has traveled back to World War II to a time when he was a twenty-six-year-old captain preparing to parachute into Normandy on D Day. Norton confirms what all the history books say about those first hours: It was all confusion.

Many of the planes carrying the troops from England were blown off course. Some of them steered away from the planned drop zones because of ground fire. Others, their pilots going into combat for the first time, got lost. Thousands of paratroopers, the spearhead of the invasion, found themselves floating down onto a darkened countryside they didn't recognize from their map studies. Miles from their drop zones, scattered, disorganized, lost, they wandered around searching for their buddies, their leaders, their targets. And as one veteran said: "Young German men with whom I had no personal quarrel were shooting at me."

Norton says they expected chaos, and they built a force that could function anyway.

"We knew that the battle was going to be won or lost by the small units," he says. "You win at the platoon level, you win the battle."

Norton's statement, true in 1944, is still true in Dave Grasso's story about "BH." Success means decision-making and action at the lowest levels.

"It starts with good soldiers," Norton says of his men. "Nothing threw them. They took what was at hand and they got the job done."

Norton looks into his drink for a moment, his white hair framed by the window and the river rolling past behind him. Sitting in this bar fifty-four years after D Day, he speaks with obvious affection about those men. "They were physically able and just fearless."

"The second thing is leadership. Those soldiers had respect for and confidence in their leaders. The leaders were the first ones out the aircraft door and into the fire."

The kind of leadership began at the top, with James Gavin, '29, who as assistant division commander (and later commander) of the 82nd Airborne Division, helped create the doctrine for employing this new kind of force. Gavin (the former enlisted soldier who inspired Cadet Greg Stitt to apply to West Point) was a Tac at West Point when Norton was a cadet and the war in Europe, already two years old, threatened to engulf the world. Cadet Jack Norton used to visit Gavin's quarters, where the officer would talk to him about tactics, about building a huge Army, about his ideas for employing an airborne force. Norton, who babysat Gavin's daughter, wound up following his mentor, and the two men fought together from 1942 to 1945. Norton ended the war as Gavin's lead officer for plans and operations.

Sitting in the bar, leaning his long frame back in the flimsy chair and peering out from behind thick glasses, Norton's bright blue eyes shine as he speaks of his combat commander.

"He set us on fire."

Gavin convinced his soldiers that they were something special. He built a unit made up completely of volunteers, from which he demanded more—in terms of performance—than other units had to give. If another infantry unit did a twenty-mile training march, Norton says, the paratroopers marched twenty-five.

The commanders gave something else to the soldiers, something that was perhaps the decisive factor on D Day. "We let the sergeants and lieutenants know, in every field exercise, in every sand table exercise, that they were the ones who were going to be making the decisions," Norton says. He holds up his hand and counts the critical points on his fingers.

"You watch 'em, you coach 'em, you trust 'em."

This gospel according to Norton is part of the Army's doctrine, promulgated in a manual called, simply, *Leadership*.

[T]he leader must let the leaders at the next level do their jobs. Practicing is kind of decentralized control in peacetime trains subordinates who will, in battle, continue to fight when the radios are jammed, when the plan falls apart, when the enemy does something unexpected.

It takes courage to operate this way . . . if subordinate leaders are to grow, their superiors must let them take risks.

And there, as anyone involved in developing leaders knows, is the rub.

Taking risks and giving new leaders a chance means those charged with developing those leaders must be willing to underwrite their inevitable mistakes. This is difficult when units (and businesses) are evaluated on numbers, when there isn't a spot on the spreadsheet to footnote, "Sure, we're off 5 percent, but we've built some leaders."

For the military, D Day provides a clear lesson on the absolute importance of pushing authority down. For most of the morning of June 6, Allied commanders weren't sure the invasion was going to work. At Omaha Beach in particular (the critical center sector of the beachhead) men and equipment piled up at the water's edge. Nothing was moving inland. The men who hadn't been killed or wounded huddled, wet and cold, in the lee of a rock shingle on the beach. Many of them had lost their weapons in the surf.

The generals were powerless, far removed. Eisenhower, the Supreme Commander, was in England, chain-smoking and pacing a hole in the floor. Omar Bradley, commanding the American forces, was on a ship in the channel; he couldn't even see what was happening because of the smoke and dust raised by the fire and bombs. Jack Norton spent his first hours in France just trying to contact his subordinate units.

The battle turned because small unit leaders each took charge of their own little part of the war. A couple of dozen captains, a few score lieutenants, a hundred sergeants all decided—independently—to do something about the mess on the beach. For each of them, cut off in his own violent little circle of the war, it looked like this: Move from here to there; press the enemy; get to the next wrinkle in the ground, the next covered position. No one man did it alone, but because they were used to being in charge, because they had been taught that leaders make decisions, these young men made individual efforts which, multiplied all along the beach and the inland drop zones, saved the invasion.

When Rob Olson tells his cadets, "You're in charge," and then backs it up (by staying out of their way and underwriting their inevitable mistakes), he is trying to instill in his cadets the same confidence Jack Norton gave his D-Day soldiers. Olson wants his cadets thinking like those soldiers on the beach: *What can I do to get this mission accomplished?*

THE END OF THE BEGINNING

A week before the end of Cadet Basic Training, the entire class of new cadets goes into camp at Lake Frederick, establishing a huge city of pup tents on the side of a low ridge some twelve miles from main post. The training schedule is a little lighter, and there is even time for some fun: games and a talent show and a few extra hours of sleep. At times it resembles a scout jamboree, except that everyone is armed with automatic weapons.

The encampment at Lake Frederick is the most relaxed time of the whole summer; at the end of the week comes the most stressful time. When they break camp, they will walk some sixteen miles on a circuitous route back to West Point, on the final foot march of Beast Barracks. Just short of main post, they will stop on the ski slope, straighten their uniforms, shine their boots, and meet West Point's band for the final two miles. They will be welcomed back by the post community, by visiting parents and, most significantly, by three times as many upperclass cadets as they've encountered all summer. For while the class of '02 camps at Lake Frederick, the yearlings, cows, and firsties all return from their summer assignments to begin the school year.

That first week is called Re-orgy Week, for the reorganization of the corps for the academic year. For the new cadets, it means they'll be transferred from their summer companies, from the squadmates and roommates they've come to know, from their squad leaders and platoon sergeants, and just about everything familiar. They'll be tossed into a new barracks, a new company, with a new and expanded cast of upperclass cadets to deal with. Like so much of the West Point experience, cadets look forward to it and dread it at the same time.

The bivouac looks like a summer camp on steroids: There are softball and frisbee games on makeshift fields; a platoon of cadets drags a big military truck up a hill in competition with a sister platoon; other cadets lounge in the sun or stand in a long line at the "boodler's," a truck that sells snacks ("boodle" in cadet slang). Beside the softball game, where the first-base coach would stand, the players' rifles are stacked in neat pyramids.

Shannon Stein's squad has some downtime. The new cadets dry out their uniforms and equipment (it rained last night while they were out on an exercise), clean their weapons, relax in the sun.

In his olive drab T-shirt, the lanky Bob Friesema looks even thinner than he did at the beginning of the summer. He talks about his night in the woods, an exercise that went by the bellicose name "Warrior Forge."

"We built a one-rope bridge about twenty hundred [8:00 P.M.], crossed the stream, then got soaked in the rain. Then I just lay in the perimeter for four or five hours in the dark. Freezing. I had no clue [what it was about]," he says, more amazed than angry. "We were just thrashing around in the dark. I didn't get to fire a round all night."

Ben Steadman, who has hurt his ankle, is worried about having to ride in on the "gimp truck" instead of getting to march back with the squad. Usually one of the most talkative members of the squad, he is relatively subdued by his status as "walking wounded."

No one wants to be left out; everyone wants to finish with the team. But by this point in the summer, they are all nursing various injuries: twisted ankles, swollen knees, rashes and blisters and scrapes. Jacque Messel has been hospitalized over the summer for a case of viral pneumonia, but she is planning to march.

The squad gathers around, all except for football player Omar Bilal. The team has already begun practicing, and the plebe football players are absent from the encampment. Although their two-a-day practices are at least as difficult as the road march, probably more, many of their classmates are suspicious. There are few sins worse than "getting over" on your classmates, that is, failing to do your share of the work, finding an easier way, getting privileges others don't have. This mind-set—and the fact that the football players miss out on the bonding experience of the road march—helps keep them separate from the rest of the corps.

Marat Daveltshin, the new cadet from Kyrgyzstan, is ready to talk about Beast in the past tense. "The training was tough, but only my English was the hardest thing."

His squadmates laugh with him about upperclass cadets asking Daveltshin the same question over and over.

" 'Are you a spy?' they ask me. 'Are you going home to lead your country in a invasion?' My army is only about ten thousand," he says, shaking his blond head.

Born in 1977, in the former Soviet Union, Daveltshin was in on the joke about the little plastic soldiers—targets—everyone called Ivan. When a cadre member asked how he felt about Ivan, Daveltshin quoted a T-shirt popular among cadets: "Kill them all and let God sort them out."

Behind the new cadets, a woman cadre member, a platoon sergeant named Lisa Landreth, is stretched out in front of her pup tent, head propped on her helmet. She wears BDU pants and a green T-shirt; her weapon is by her side, and she is reading *Mademoiselle* magazine. The model on the cover, all slinky dress and sultry pose, is from some other planet.

The new cadets are eager to talk and, for the first time all summer, not in the same uniform. Some of them wear BDUs, some wear PT gear, a few of them have removed their BDU blouses and stand in their green T-shirts.

Clint Knox says the two squad leaders they had this summer had completely different leadership styles. "I guess it's good to get used to different kinds of leaders," he says. "[First detail squad leader Grady] Jett taught us step by step, showed us how to get things done. Cadet Stein enforced standards, but she let us figure out a lot of things on our own. Maybe she was taught that way."

Pete Haglin adds, "I learned you should always keep your people informed."

Some of the eagerness has gone from his voice. All the new cadets are concerned about the looming threat of Re-orgy Week, and Haglin carries additional fears about academics. But for him, there is something more: He has come to resent Shannon Stein.

"During the second detail," Haglin continues, "half the time I didn't know what was going on. Then they [the cadre] get frustrated and take it out on us. I liked Jett better. If something got messed up he'd be totally honest about it. Second detail we did what we were told and we were still screwed up."

Lamb adds, "I liked Jett better, but Stein knew us better."

"I learned never turn your back on your squad," Knox says. "She turned her back on us. It's one thing to have high standards, it's another to tell someone, 'You're worthless.' "

The new cadets are on a roll, and they aren't shy about voicing their complaints.

"When she didn't like doing things she made me do them," Steadman says.

Barry DeGrazio says that Steadman became the "assistant squad leader." It's not an official position, of course, but most of the squad leaders find a sharp new cadet to help things move smoothly.

"She didn't like to get up before everyone so she'd yell at me to get everyone else up," Steadman says. "I follow orders and all that, but it's tough because your classmates get pissed off at you."

He pauses for a moment, perhaps considering how he sounds, having said those things out loud. "I guess I'm learning to be a leader, too," he begrudges her.

"I know I won't yell at my plebes or use profanity when I'm an upperclassman," DeGrazio says. "When I messed up during first detail, I felt bad. When I messed up second detail Stein yelled at me all the time. After a while, it didn't matter. It was like the boy who cried 'Wolf!' "

Lamb hated the false motivation, "Whenever we went anywhere, we'd yell and have to say 'Hoo-ah.' It got stupid after a while."

Pete Lisowski says the most important thing a leader can do is know and understand his people. "Jett came back to check on us before we came out to Frederick. [First detail platoon sergeant Greg] Stitt came out, too. They cared about us and wanted to know how we were doing."

"Jett admitted it when he made a mistake," DeGrazio says.

"Stein didn't," Haglin says. "She'd pretend she didn't do anything wrong."

Bob Friesema thinks before he speaks. The summer's experiences are still so fresh that it's hard for him to get a handle on what he's seen. "The best leaders are unselfish," he says. "They care more about their troops than they care about how they look to the people above them."

"They make it tough because they want us to be tough," DeGrazio says. "Basic is supposed to weed out people who can't make it." He all but parrots the Commandant when he adds, "We've learned a lot about being soldiers and a little about being cadets."

Daveltshin thinks in Russian, then translates to English. It takes him a moment to join any conversation. "I saw many different styles; each cadre has his own."

Stein, they agree, is all about authority. But she undercut herself because she bucked authority: when she flaunted Mess Hall rules, when she was in open disagreement with the platoon sergeant and platoon leader. First detail platoon sergeant Greg Stitt, the serious former helicopter crew chief, gets the highest marks.

"He was great," they agree. DeGrazio elaborates. "It was based on respect. If you made a mistake you felt bad because you let him down. He didn't have to yell."

DeGrazio recalls a time when Stitt dropped eight new cadets for push-ups over some infraction. When the whole platoon got down and started doing push-ups, the platoon sergeant was pleased. "That's what he was looking for, but he didn't cut us any slack. We did push-ups and flutter kicks for a long time."

"He helped me e-mail home," Daveltshin says. Then he recalls Stitt in the gas chamber, mask off, doing jumping jacks and reciting plebe knowledge.

"He did what I cannot do," Daveltshin says. "I would feel safe with him in combat. A soldier has to be tough because combat is the toughest situation in the world, the most stressful. He has to be able to handle that stress to save his life, to save his buddy's life, to win the battle."

"When the leader is good," Steadman says, "you want to do well."

Not every leader they encountered garnered this kind of respect. The new cadets complain about personal attacks, jokes at their expense. One claims that when Stein's boyfriend, a second class cadre member from another company, visited her in the barracks, he and Stein amused themselves by teasing the new cadets.

"She called people homosexuals and assholes," Lamb says.

Many new cadets are required to memorize, and recite on command, special "poop," a little entertainment for the upperclass cadets. Often poop is dreamt up to go with a name, a hometown, a special talent. A new cadet named Springsteen had to know a repertoire of Bruce Springsteen's songs that she could sing on command.

Lamb's poop, which he bristles at now, is, "Ma'am, I am a product of my parents' sick sexual fantasies about farm animals."

Stein told Friesema to pick flowers, and she mispronounced his name so that it sounded like a popular woman's fragrance. The upperclass cadets thought the jokes funny, but the humor wore thin fast for the plebes.

The hierarchy of leadership they describe is similar to one described by Gus Lee, the author of *Honor and Duty*.

"We have three powers to get people to do things they otherwise might not," Lee said in an interview. "The first power is authority: Do this, or I'll make you walk funny. The second power is motivation: Do this and I'll reward you. The third power is inspiration: Follow me."

This "inspirational leadership," to use the jargon popular at West Point, is the most difficult. Leadership, in this case, means getting people to do what the leader wants because they, the followers, want to do it.

Grady Jett and Greg Stitt hit this mark. The new cadets "felt bad" when they "messed up" under these first detail leaders. With Stein, they just got yelled at.

A subdued Shannon Stein sits by her pup tent. She has just taken a cold-water shower (the only kind available) and is scrubbed clean; her uniform is clean, her hair still wet. She knows the new cadets are not very happy, and she cares. "It's hard for them to understand [how difficult the job is] until they've been in a position of responsibility," she says.

Stein just received her military grade for the detail. (Military grades are factored with academic and physical performance to determine class rank, which in turn influences a senior's choice of post-graduation assignments.) The platoon sergeant and platoon leader she butted heads with all summer gave her a C.

"I could have put myself first, kissed up to the chain of command, but I put [the squad] first and myself last." She picks a blade of grass and stretches it between her fingers. "They were a great squad, and I had a great time. I told them all in counseling: Don't compromise your integrity."

She considers what she might do differently, given another chance. "I needed to pay more attention to them individually. They passed off [recited] knowledge together. Some were better than others; I focused on the team. Re-orgy week will be hard because it's unfamiliar," she allows, perhaps thinking of the same table duties that worry her charges.

"You don't get too much praise in this position," she says sadly.

But she did learn a lot about her own capabilities, and about the importance of leading by example. "You can't just talk about leading by example. Like when we were on the Confidence Obstacle Course."

As the name states clearly, this course is all about facing and overcoming fear. Cadets climb over a series of high obstacles: towers and platforms and thirty-foot ladders with rungs made of telephone poles.

"I could do it—with apprehension—but I did it in front of them. I had to get Daveltshin down off one obstacle. He kind of froze up. So I'm up there, all five three of me, talking him down and showing him how to do it."

Although only two years older than the new cadets, Stein is able to see beyond the surface conditions they gripe about. She has her own analysis for what was going on in the squad.

Of Pete Haglin, whose attitude changed the most—for the worst—over the summer, Stein says, "Haglin was my problem child, always making excuses. He was this high school stud. Teachers loved him; he was popular. Then he gets here and he's none of that. He's also trying to live up to his father. He wants his father to be proud of him. He came into my room bawling one night because he thought he was a failure. I had to give him 'the talk,' she says. " 'This is West Point; whatever you do, it's not going to be right.' " She doesn't smile as she says this.

If her insight on Haglin is on-target, she's missed the mark with Jacque Messel. Messel, whose performance has been consistently strong throughout the summer, is all but convinced that West Point is not for her. She is still here, a week from the beginning of classes, only because it is her father's dream.

"She improved tremendously," Stein says of Messel. That much is true, but Stein has misjudged what it means. "She's definitely staying."

Stein sets her cap on her head, pulls it low, and heads down the long row of tents to where she has asked the squad to gather. They sit on an embankment shielded by a little hill as Stein paces the dirt road in front of them. She is taking a risk, and she knows it.

"I was sitting in this exact spot two years ago," Stein tells them. "And in two years I'll be out in the Army."

She pauses, as if thinking how much work there is to do in a short time. "My main goal this summer was this: I want you to walk out of here with self-confidence, with the attitude, 'I can handle anything they throw at me.'"

She claims that is the point of the squad's odd motto, "Always the Hard Way."

"I love you all the same and I hate you all the same," she says, capturing in one phrase an essential truth about the cadet experience at West Point: It's a love-hate relationship.

She promises them a pizza party one Friday evening when they have free time.

"You'll be hungry by then. I've been a candy-ass at the table, and you're going to catch hell at your new tables. But the stuff I put you through mentally was tougher than all that other stuff."

The new cadets have been quiet up to this point. They watch her, and they look past her to the ridge they'll climb in the morning to start their march back to West Point and Re-orgy Week.

"I want feedback too," she says. "This was my first time in a leadership position."

Clint Knox says, "On that first day, you were the only one who knew stuff; the platoon sergeant was lost."

"You guys are like my babies," Stein says.

Ben Steadman asks, "Were you as scared as we were, ma'am?"

She tugs her hat even lower, posturing. "Whaddya think?"

"It was easier when I knew what you wanted," Pete Lisowski ventures. She is still an upperclass cadet, still their squad leader.

Haglin, who has become the squad's malcontent, says, "I'm still trying to figure it out."

Stein smiles nervously. She asked for this, but she isn't enjoying it.

Jacque Messel adds quickly, "You were tough when you needed to be, but you let us relax."

"You'd take us up and down," Lamb says.

"What's the purpose of Beast?" Stein asks. Then, in another pronouncement that is a part of cadet lore but certainly isn't part of the official Academy position, she adds, "It's to break you down, then build you back up."

Daveltshin smiles and says, "I was proud to be in the most popular squad in the company. All our other classmates were afraid of you."

Stein does have a war face, in spite of the fact that it's perched on a diminutive frame. Friesema, who has been watching quietly, speaks.

"With all respect, ma'am," he says, mustering his courage, "We're usually clueless. You think the questions we ask are stupid, but we really don't know the answer."

"Maybe I could have listened better," Stein says quickly.

"Yelling doesn't work," Haglin adds. "I've been yelled at so much it has almost no effect."

Stein glances at Haglin, stands her ground in the dusty road.

"You remind me of my mother," Tom Lamb says. The squad laughs, releasing some of the tension.

"You're the same height, have the same color hair. She even yells at me the same way. I've been dealing with it for nineteen years, so it wasn't really effective."

Knox continues the levity.

"Before we came here, we were all like, 'I'm the man.' Before I came here I thought I was God's gift, you know?" He smiles broadly. "Now I think God made a mistake."

The new cadets fidget, shift their positions on the grass. Whatever other feedback they may have for their squad leader, they aren't sharing it.

"OK," Stein says. "I love you all; I wish you the best. I'll be checking on you. I'll write to your parents and tell them how you did."

Nervous laughter, a few groans.

"There'll be more good points than bad points," she says. Then, "Bring it in one last time."

They huddle for a cheer. "Always the hard way, ma'am."

Later that afternoon, Deborah Welle and Jacque Messel sit side by side in front of their shared tent. Welle, who has injured her ankle, wears a plastic brace she calls her "Robo-Cop boot." She won't make the final road march.

Messel, who is fair-skinned to begin with, looks pale. In the last two weeks she has missed five days of training and has spent several days hospitalized. She also feels as if Stein has already written her off as a washout.

One day, shortly after returning from the hospital and a round of treatment for the virus that has been plaguing her, Messel felt as if she was going to be sick during lunch. The medication made her nauseous. When she asked Stein if she could be excused from the table, Stein asked her, in front of the squad, if she was bulimic. When Messel said no, Stein asked, "Are you lying to me?" The fact that her squad leader had questioned her integrity—a serious thing at West Point—upset Messel quite a bit.

Now, Messel looks tired. There are circles under her eyes, and she moves and speaks slowly. Her rifle lays across her lap, open at the hinge for cleaning. "I have to make up a lot of training," she says. "But I'm going to do the road march."

The march is sixteen-plus miles, with a field pack and weapon, helmet, and boots. Although the lead companies will step off long before dawn in the morning, the new cadets will spend hours trudging along in the sun. It's not much of a prescription for recovering from viral pneumonia.

"It was right to stay for Beast," she says, "but I'm still pretty convinced that this place isn't for me."

Messel pulls the bolt from her rifle and wipes it with a rag, turning it over in her hand. Then she snaps it sharply so that it will seat when she puts it back in the weapon; she handles it as if she's been cleaning M16s all her life. "I talked to my dad about wanting to leave. He's coming out to see me this weekend so we can talk about it."

Although it will be her first visitor from home, she is not looking forward to it.

"It was always his dream that I come here," she says. And suddenly her eyes sparkle silver with tears. She slaps her weapon shut and, in a steady voice, says, "I'm just ready to get on with my life, start making other plans."

On the last evening of Beast Barracks, the new cadets put on a talent show. Lieutenant General Dan Christman, '65, the Superintendent, brings the Chief of Staff of the Army, General Dennis Reimer, '62, as his guest. Reimer has made the last three shows here, and has marched back to West Point with each class.

"You may have heard that I always start with the last company," Reimer says to the fifteen hundred cadets and guests gathered on the hillside. "And I try to catch up with the lead company by the time we reach West Point. Last year, I didn't quite make it."

He pauses. "The fellow carrying me just couldn't make those last few yards. This year, though, I've been assured that the cadre will do better."

Everyone laughs at the general's joke and the show gets under way. The night is comfortably warm, the setting spectacular: The stage is downhill from the encampment, and the audience can look beyond it to wooded ridges in the blue distance. Above the hills, backlit clouds stand like God's watchtowers. The cadets sit cross-legged on the ground, grouped around the little company flags called guidons. They are fed and relaxed and able, for a few minutes, to forget about what will unfold for them long before dawn the next day. This is one of those times when the West Point experience sits somewhere between Army basic training and freshman orientation at some mid-sized college.

As the show begins, the first sounds don't come from the stage at all, but from the field behind the crowd. Everyone turns to see a bagpiper, a new cadet in BDUs, his rifle slung diagonally across his back. He marches down the aisles formed by his classmates, and the weird skirl floats over the heads of the crowd and rolls around the valley. The cadets applaud wildly.

The first act is a group of new cadets singing gospel. Any practice time they've had was stolen from a packed training schedule, but they choose simple arrangements and pull it off beautifully. Next, three men in BDU pants and green T-shirts do an elaborate dance of robotic hip-hop and foot-stomping. Then a young woman takes the microphone and dedicates her song to the "2 percent club," that tiny group of cadets who keep the same boyfriend or girlfriend back home over four years. Her face is hidden by the BDU cap, but she has a pretty voice and people in the crowd sing along.

Fourth squad's Clint Knox, who graduated from a military high school, takes the stage with a drill team from Alpha Company. They spin rifles in an elaborate ballet onstage; in a solo performance, Knox spins a rifle in each hand. The final act is also from Alpha Company. A chorus that includes Jacque Messel takes the stage and sings Lee Greenwood's "God Bless the USA."

West Point has something of a love affair with this treacly song. A former Superintendent, Dave Palmer, played it during assemblies with cadets, staff, and faculty. Its lyrics can be heard at meetings, sports events, and rallies.

Yet here, amid all these young faces and earnest singing, it seems appropriate. They have plenty of time to become jaded and cynical. Tonight it is enough that they are with friends, that they are at the end of basic training, their first major test. The hillside is a great, green sprawl of possibility and youthful enthusiasm. And if any of them bother to look up, they can see that they have taken on a thousand allies, that they have become a part of something larger than just themselves. Some of them will revel in that new identity, while some will be consumed by it.

Later, the new cadets gather near their tents as dusk drops around them. Flashlights bounce through the tent streets.

"I get a sick feeling when I think about Re-orgy week," Friesema admits.

They worry about small things that will loom large tomorrow when they are faced with a company full of unfamiliar upperclass cadets. Is their hat brass shiny enough? Their shoes? Do they really know all the songs, cheers, West Point history, Army history and factoids well enough to make it through the next few days without gaining a reputation as a screwup? Even as they worry about what they don't know, they're amazed at how far they've come.

"I think I should look like I did before, then I look in the mirror and see this stranger, with this haircut and these glasses," Barry DeGrazio says as he runs his hands over his buzz cut. He wears the strap-on military-issue glasses that the upperclass cadets call "BCGs," for "birth control glasses." As in, "Those glasses are so ugly you could never get laid."

"When we got here we didn't know anything. There was so much expected of us and it was all new. Some of the mistakes I made just made me feel stupid."

Friesema tells a story about trying to find a way to carry his retainer to meals. "When we wear white over gray, there aren't any pockets," he says. "And I'm supposed to wear this stupid retainer. So I put it in the side of my garrison cap."

The garrison cap, also called a fore-and-aft cap, fits on top of the head like a gray envelope.

"By the time I got down to formation, it had worked itself out and was just hanging on my hat. Cadet Jett got a look at it and couldn't believe it. I had about ten upperclassmen around me laughing at me and yelling at me at the same time."

Friesema smiles, and his classmates laugh with him. They are all amazed at the changes they've gone through.

"I used to be pretty undisciplined," Friesema says. "I liked to just sit and watch TV. In my whole senior year I bet I did one hour of homework. I never cleaned my room, either." His surprise is not that he gave all that up, but that he used to live that way in the first place.

"If someone told me to do something, I'd work hard to get out of it," he continues. "Now I do what I'm told and do it as well as I can."

Pete Lisowski notices another change. "The other day I said 'Hoo-ah' on the phone with my mom," he says in his slight North Carolina drawl. His squadmates laugh.

"They said they were bringing up Bojangles chicken [on the upcoming weekend visit] and I said 'Hoo-ah!' They live near a Marine base and have heard this stuff. She asked if that's like 'U-rah,' and I said, 'Yes, I guess so.' She said, 'I don't want to hear it.' But my biggest change is that I used to take charge of everything. It had to be my way . . . or everyone would know that I didn't agree. Now I'm more of a team player."

"I appreciate the small things," Friesema says. "I would kill to be able to walk up to a refrigerator and grab what I wanted."

"My first phone call, I was all choked up," DeGrazio adds. Then, not wanting to explore that area too deeply, "The big change for me is that I used to be a musician. I play bass, tuba, any kind of guitar. Now I'm someone with no music, except cadence calls, I guess."

"I miss music, too," Friesema says. The new cadets are not allowed radios or CD players. To the list of depravations, Friesema adds the great American icon of independence.

"I miss driving. If you got bored you could just hop in a car and go to the mall, to a friend's house, to the gym." Except for leave periods, that privilege is far away. Cadets are not allowed to have cars until near the end of junior year.

Daveltshin, who is four years older than his squadmates and twelve thousand miles from home, says simply, "I miss my girl-friend."

The sun is gone, though it is still light enough to see figures moving through the tightly packed rows of tents. The bagpiper plays somewhere nearby, and the air is wet with threatened rain. Many of the new cadets are expecting their first visitors over the coming weekend. Friesema's father, two grandfathers, and two younger brothers are coming from Wisconsin. Lisowski's family is driving up from North Carolina, with the promised bucket of Bojangles chicken.

The new cadets consider what their parents will see different in them. They are quiet for a few moments, and in the darkness it's hard to see if they're considering their answers or just daydreaming about home.

"I think we're all a lot more mature," Friesema says. "I've grown up more in the last six weeks than in the first eighteen years of my life. . . . I'm amazed at the total authority the upper class have. I mean, your parents had a lot of control, but at least you had an opinion. Not here."

"I didn't have any problem with that," Daveltshin says. "This is the only way to train soldiers. It develops immediate discipline so you can handle pressure. It doesn't offend me that cadre is younger than me; they know more than I do."

Shortly after dawn, Alpha Company pulls into a rest stop. Their morning started at 0330; they are already sweat-soaked, though the weather is mercifully cool for a Hudson Valley August. The march is orchestrated to the tiniest detail, with down-to-the-minute timetables for rest halts, road-crossings, checkpoints. The purpose is to keep the new cadet companies from piling up on one another, especially as they cross the busy roads that run along the valley floor.

Most of the sixteen-plus miles are done on dirt roads that wind through the woods. It is humid under the trees, and the new cadets walk in silence, one file on each side of the trail. General Reimer, the Army Chief of Staff, moves up in the space between the files, greeting the new cadets he passes. He is tall, six four or so, and trim, and his long legs eat up the distance. He is working his way from the rear of the column, so most of the new cadets don't see him coming, and although they are polite and respectful, the ones who do speak to him are hardly in awe.

The trail winds down the eastern side of a ridge, the sun poking through the trees; it's less than half a mile to Round Pond and a rest stop. Jacque Messel, rucksack square on her shoulders, weapon at sling-arms, moves along at a good clip, keeping pace with her squad. Suddenly her breathing becomes labored. She slows, stumbles a bit, then steps off to the side and bends over to put her hands on her knees. Her breath comes in short, loud gasps. Her squad calls some encouragement to her and a cadre member points to the top of the next hill—so close—but no one slows down. When the company medic catches up with her, he loads her on a vehicle to drive the few hundred yards to the rest stop. She is suddenly faced with the prospect that she won't get credit for completing the whole march, and she'll wind up with an "incomplete." As the medic walks her to the vehicle, she wears a look somewhere between disgust and fear.

Round Pond is one of the recreation areas that dot the reservation at West Point. A ring of campsites encircles the lake, and there is a stone- and wood-building that's used for parties. Strong morning light comes in at a low angle. A half dozen new cadets who are injured and can't make the march hand out apples, bagels, and Gatorade as the squads come in off the hillside trail. Alpha Company moves in as the company ahead prepares to move out. There is a great deal of shouting as the arriving new cadets are shuffled into the shade and the departing platoons are moved onto the road.

"You got ten minutes to rest," Alpha Company's first sergeant shouts. "Make sure you fill up your canteens and use the latrine."

Shannon Stein's squad sits in a tight clump. There is plenty of room, but no one has told them to spread out. Some of them pull their boots off; they wear black dress socks beneath their green GI socks to reduce the friction. The cadre shout instructions about filling canteens and checking for blisters, but none of the squad leaders circulate and check that their instructions are carried out.

DeGrazio isn't worried about his feet; he's worried about his shirt, which has a small hole in the back, right between his shoulder blades. He does not want to go into his new company with a hole in his uniform. He has packed an extra shirt inside his rucksack and wonders if he'll have time to change at the last rest stop.

Daveltshin shuffles as he walks. He has a painful heat rash between his legs, and every step rubs his already raw skin. Someone points out that GI footpowder might help; a cadre member from another platoon finds some for him. Messel rejoins the squad before they finish their Gatorade. She has missed only three hundred meters of the march and wants to go at it again. Stein calls to her from where she's sitting. "Are you marching?"

Messel isn't sure how to answer; no one has told her what to do either way. Maybe Stein is asking for her opinion. "I guess so, ma'am," she says.

"Get up at the front of the squad when we start out, then," Stein says. "DeGrazio, push her if you have to."

Messel has pulled her weight all summer, so she is not ostracized; but she has been talking about resigning since the first detail, so she is not completely a part of the group. In these last hours of Beast, she is drifting away, like a relation who hasn't called in a while.

As they get closer to West Point, yesterday's exuberant new cadets become quieter. The cadre members, who are about to be relieved of their responsibilities, are happier with each passing mile. Shannon Stein is a little apprehensive about the start of soccer practice. She faces two hours of running that very afternoon. But she is looking forward to the academic year because she gets to "sleep in until 5:45."

"Saddle up!"

The command brings some groans, and a cadre member reminds them of one of Greg Stitt's favorite expressions: "Pain is just weakness leaving the body!"

The platoons form up on the unpaved road, then wind uphill, one file on each side, ten meters between new cadets. Friesema marches holding his rifle in one hand and a sweaty, tightly folded piece of paper in the other. "I'm memorizing 'The Days,' " he says.

"The Days" is a bit of plebe trivia the new cadets are responsible for, a list of a dozen or so events for the coming year and the number of days remaining until each. Each sentence must be recited exactly as it is spelled out in the plebe knowledge booklet. No contractions, no word substitutions, no missteps or mispronunciations, nothing left out.

"Sir, there are fifty-seven and a butt days until Army beats the hell out of Navy in football at Veterans Stadium in Philadelphia!"

Because the list is long—it includes major sporting events, Ring Weekend for the first class, Christmas leave, spring leave, Graduation Week—and because the numbers change every day, it is difficult to keep straight. That makes it one of the more popular pieces of plebe knowledge the upperclass cadets will demand. And Friesema knows that those upperclass cadets—all of them strangers—wait by the hundreds back at West Point. As he has done all summer, he turns his anxiety into action, memorizing the numbers as he marches.

The new cadets have been told that the first impression they make today, their first day in the new company, will likely stay with them for the rest of plebe year. A new cadet who gets a reputation as a screwup in the first few hours after the march back from Lake Frederick is simply going to have a tougher year than one who does well. Part of the elusive "doing well" means knowing fourth class knowledge, and knowing it cold. As part of that store of knowledge and skills, the new cadets are supposed to know how to perform the intricate plebe table duties. Stein's squad does not feel prepared for that test. Friesema isn't going to get caught short on anything else. He looks around, notices he is the only one studying. "I guess I'm pretty nervous," he says.

Just ahead of Friesema, Stein marches at the head of her squad, dark eyes fixed straight ahead. She is a small woman, and the frame of the rucksack extends from above her shoulders to below her waist. Her legs work like pistons as she tears at the uphill slope; she has trouble on the down slope because the rucksack, which weighs nearly half what she does, threatens to send her flying downhill. She keeps her eyes on the ground a steady three feet in front of her as she considers this leadership lab that's about to come to an end. "You have to be confident and intelligent," she says between deep breaths. "You have to set the example so people will want to emulate you." *Pant, pant.* "You have to work as a team."

Suddenly Stein is passing a new cadet who is falling behind a squad farther up the line. Her helmet is sideways on her head, and she looks as if she's about to cry.

"Let's go, Reece!" a cadre member shouts from somewhere ahead. But Reece is downshifting on the big hill; her squad is pulling away from her. Suddenly her platoon sergeant comes jogging back. It is Lisa Landreth, the second class who spent part of yesterday afternoon reading *Mademoiselle* outside her tent. "Let's go, let's go, let's go!" Landreth yells. "One more hill until the ski slope!"

Reece makes an effort, manages to keep it going. Landreth runs forward again for forty or fifty yards, the length of the platoon, to check on the lead squads. She is not much bigger than Stein, and as she runs, her helmet and rucksack bang up and down.

Two minutes later and Landreth is back, still running, still being pummeled by her equipment. She carries her rifle at port-arms, holding it diagonally across her chest. Winded, she still manages to shout encouragement. "Let's go, Reece! Let's go!"

Colonel Joe Adamczyk, the Brigade Tactical Officer, catches up to Stein. He is unfailingly polite and even friendly, but he has a reputation among the cadets as something of a martinet, a nit-picker who is fond of rules and regulations. He is thin, like an overworked long-distance runner. Adamczyk also served as a company tactical officer in the early 1990s. He has a lot of experience on the ground at West Point, and has also commanded an infantry battalion and served in a variety of other jobs around the world.

"When I came back here as a Tac and realized all the things we used to do wrong, especially to plebes, I was embarrassed," he says. "All that yelling and hazing; you just can't treat people like that. And we weren't teaching good lessons to the upperclass cadets, either."

When Adamczyk leaves, Stein says, "What a bunch of bullshit. He's always yelling at cadets."

At mid-morning the Class of 2002 starts assembling on West Point's ski slope, just outside Washington Gate. They sit in the bright sunlight and congratulate themselves. There are only two miles to the cadet area and the end of Beast.

Barry DeGrazio pulls his clean BDU shirt from his rucksack and puts it on; the one with the hole in the back goes into the ruck. He doesn't look relieved. Bob Friesema and a few others frantically study their plebe knowledge. They take turns reciting the "General Orders" for guards. A few cadets have brushes and shoe polish, which they use to clean their boots.

On the ski slope, two figures in full gear move straight uphill. One is second class cadet Brad Marvin, the company first sergeant; the other is Jacque Messel. She is only two or three hundred yards short of doing the same march her classmates did, and she wants to get credit. Marvin invited her to tackle the steep slope, and she took the challenge. By the time the two reach the bottom, they are both completely drenched in sweat. Messel smiles like a lottery winner.

Alisha Bryan, the Alpha Company "milk and cookies lady," is happy that Messel made it through Beast. "When you go through something like this, you can look back and see how far you've come. That helps you get through whatever is next."

She looks out over the new cadets in the few minutes of quiet celebration allowed them. "They're a little cocky. Re-orgy week will take that out of them."

The new cadet companies form for the last leg of the road march, the parade onto West Point proper. They march in a tightly packed formation; a cadre member walks alongside, calling cadence, and the new cadets join in the marching songs.

My girl has a wooden leg
That's why I call her Peg
I'd buy her anything
To keep her in style!

The companies squeeze through the gate at the top of Washington Road, then down the hill in front of the Keller Army Hospital, where the USMA Band waits to join them. As they pass the first sets of quarters, the cadets can see whole families turned out to watch. The parade is a major event for the community. It marks the clear line between summer activities and the school year. (The calendar of the West Point elementary school, an on-post school for the children of staff and faculty, coincides with the cadet school year. Everyone starts at once.)

The adults along the route sit in lawn chairs or stand talking to neighbors while the children play on the manicured grass. Near the teen center, three little boys hug a large tree beside the road. Their faces are painted in black, gold, and gray, and they hold a hand-lettered sign, "Welcome '02!"

Now the band plays, and the cadence callers lower their voices to an almost conversational tone. Left, right, left, right, left. In ranks, the new cadets keep head and eyes to the front. With their helmets pulled low, their faces and uniforms sweat-soaked, rifles at sling-arms, they are reminiscent of newsreel photos of World War II GIs on parade in Paris or Rome. The formations pass the cemetery, with its concentric rings of white headstones, its several dozen outsize monuments: an obelisk for George Custer, a football-shaped stone for Army coach Red Blaik.

Closer to the cadet area, families of new cadets stand on the curbs and sidewalks, checking the guidons for the company letter, then scanning the ranks for a familiar face. Some of the families have been warned, by others in their local parents' clubs, that they shouldn't put the names of their new cadets on the signs. No new cadet wants to be singled out, especially on this day, when battalions of upperclass cadets await them.

"Peoria, We're Proud of You!" one sign says. The parents holding it are decked out in West Point hats and T-shirts.

"Welcome Back, Ohio Buck!"

Another sign seems designed for the whole class. "I'm so-o-o proud of you! Love, Mom."

One family, from Syracuse, New York, is dressed in the orange of Syracuse University. "We told him what to look for when we talked to him on the phone," the mother says. There are two little sisters in orange T-shirts, one with a huge foam rubber hand (also orange) in a "#1" gesture. The other sister holds a sign with "#5" on it; it was her brother's number when he played sports in high school. Mom has an American flag. Dad is busily video-taping everything. Nearby, another family from neighboring New Jersey has turned out in force: Besides the parents, there are two grandmothers, several aunts and uncles, and some neighbors from home. Scores of officers from the faculty stand along the route, also. The West Point graduates in the group talk about their memories of the march back from Lake Frederick.

Two new cadets march in front of the regiment, holding a long banner with the class motto: "Pride in All We Do: 2002." Behind the banner, General Reimer, the Chief of Staff, and the Superintendent are in the front rank of officers.

As the new cadets turn the corner in front of the Dean's quarters, they can see ahead of them a wing of MacArthur Barracks and a dark sally port leading to North Area. On this short stretch of road in front of Quarters 100, the crowd changes. The civilians give way to uniforms, and just past the reviewing stand both sides of the road are lined with upperclass cadets.

All summer long the ratio has been three new cadets for each upperclass cadet. And yet there were enough cadre around to make the new cadets feel as if they were always being watched (they were), always being corrected (ditto), always doing something wrong (definitely). Suddenly there are three times as many fault-finders. The sidewalks are full of them; the road is lined with them. They smile because they know what's about to happen. The new cadets only know it's going to be unpleasant.

The cadre has been talking about this day all summer long. Most of the stories have been in the tradition of the boogeyman tales used to scare children around the camp fire. Under the stars at Lake Frederick, surrounded by classmates, the new cadets could still joke about it. No one is laughing now.

"Why aren't they smiling?" one of the grandmothers asks. "I mean, they're finished, you'd think they'd be happy."

Reimer and Christman mount the temporary reviewing stand in front of the Superintendent's quarters. From there they return the salutes of the passing companies.

The new cadets companies form on the apron—the same spot they've been occupying all summer—ground their gear, and head into the Mess Hall for their last meal in Beast Barracks. When the new cadets disappear into the big doors of Washington Hall, upperclass cadets in razor-sharp white-over-gray uniforms gather on the grass in front of the phantom formations of gear left behind by the new cadets. They plant the guidons of the thirty-two lettered companies (the organization of the corps for the school year) in a long line that stretches around the edge of the concrete apron. There are a half dozen upperclass cadets from each company, come to collect their contingents of new cadets. By the time the Class of '02 comes out of the front doors of the Mess Hall, the wide stretch of green that had been empty fifteen minutes earlier is now filled with upperclass cadets at rigid attention. It is a bit of theater, heavy-handed but effective.

"Let's go, new cadets. Move out, move out, move out!" The Beast cadre shouts. They are eager to turn over their charges, let them be someone else's problem. Let someone else worry about getting their weapons clean and turned in, their gear moved to new barracks and stowed.

"MOVE, MOVE, MOVE, *MOVE MOVE MOVE!*"

The new cadets hurry down the big steps of the Mess Hall and scurry to their equipment. Over in Alpha Company, the platoon sergeants shout gleefully, "Get it on!" The new cadets strap on their helmets and equipment, then hold their weapons at their sides. In front of them they can see the silent cadets in white-over-gray, standing at attention. Waiting.

Bob Friesema stands in the back rank of his platoon. His notes and plebe handbook are tucked away in his rucksack or in his pockets. He knows what company he'll be joining, and he has already moved some of his gear. But there is a long day ahead of him, when he'll jump to the commands of a new chain of command.

The new cadets come to attention, and there is a silent moment as the two sides face each other across a few yards of empty concrete. The families on diagonal walk-shuffle nervously. Friesema chews his lip.

The commanders of the CBT companies turn to their new cadets, who sound off with their company mottoes. Then the firsties give their last instructions of the summer.

"When I call your company, fall out and fall in on your new company guidon."

"Alpha one!"

"Bravo one!"

"Charlie one!"

As each company is called, new cadets assigned to that company lift their M16 rifles to port arms, step out of ranks, and run to the guidon of the new company.

"Delta one!"

"Echo one!"

During the academic year, the Corps of Cadets is organized into a brigade of four regiments. Each regiment has eight lettered companies. Companies A-1 through H-1 make up the first regiment, then A-2 to H-2, and so on through to the last company of the last regiment, H-4. New cadets are not assigned at random; Academy officials use a computer program to give each company a cross section based on academic performances, varsity sports participation, race, and gender.

As the new cadets run toward their new company guidons, the upperclass cadets in white-over-gray begin shouting orders. "Fall in! In a *file,* new cadet! You know what a file is? What have you been doing all summer? I can tell you right now *the vacation is over!*"

The white shirts circle the sweaty new cadets, correcting their posture, criticizing their uniforms, finding fault with the way they stand at attention, the way they hold their rifles.

Bob Friesema is one of the last to leave his squad. He stands at attention as he is, quite literally, abandoned by the team he has worked with all summer. He admitted to being nervous about what the next few days hold, and it is all visible on his face. When his company is called, he brings his rifle up smartly and runs to meet his next challenge.

When the new cadets fall in behind the guidons of their new

companies, cadet basic training is over. During the few days that remain before the beginning of classes, they will be issued books and computers; they will receive schedules and classroom assignments. The luckiest among them will have some third class team leader march them into each academic building to show them how to find their classrooms.

They may even spend a few minutes reflecting on what they have accomplished in the few short weeks since R-Day. But they won't pause; they're on a fast-moving train that isn't about to slow down.

LEARNING TO FOLLOW

The smothering humidity of August gives way to a mild September and a crisp October as the academic year gets rolling. From the parking lot on the roof of Thayer Hall, it's easy to know why so many artists come to this valley. The Hudson changes costume many times in a day, with each subtle shift in the light. It knows dozens of blues and grays. Its surface is often ice-choked in winter; a strong wind coming down the valley can whip up whitecaps; a full moon makes it look silver-plated.

Cadet life changes from the summer's emphasis on military requirements to the bookish demands of the fall. The cadets' world shrinks, too. Upperclassmen return from far-flung summer assignments with the Army to spend their days shuffling the short distances between the academic buildings, the gymnasium, the Mess Hall, the barracks, and the chapels.

At 6:00 on an October morning, Pershing Barracks, with its tower, gargoyles, and crenellated roofline, looms out of the darkness like a dream of some castle. Its big windows are still mostly black; the few lights clicked on are in plebe rooms. Just inside the enormous oak doors stands Cadet Third Class Barry Huston, who is already dressed in his class uniform. Huston is Bob Friesema's team leader, the first line supervisor in a long chain of command. He is responsible for teaching Friesema much of what the plebe needs to know to get through this first year. It was Huston who, during Re-orgy Week, showed the plebe where his classrooms were; who took him to issue points for his books and computer; who showed him how to wear the parade uniform of full dress gray, with its spiderweb of white belts. Huston monitors Friesema's grades, his command of fourth class knowledge (the requirements change every month), and his duties in the company (such as delivering dry cleaning and laundry to the upperclass cadet rooms).

Huston heads up the wide staircase with its ornate banister. All around, cadets are stirring. Some are dressed in gym gear, walking to the showers, towels and razors and soap in hand. A woman in a long cadet bathrobe shuffles from the latrine, her shower shoes squeaking on the shiny floor, her hair dripping. Huston approaches a door; one of the paper name-tags says, "FRIESEMA R D 02."

West Point even teaches cadets how to knock: two sharp raps. Here a polite tap is the equivalent of a weak handshake.

Friesema answers the knock by shouting, "Enter, sir!"

Huston looks in the room, then retreats. Friesema goes back to his morning rituals. He has lost the deer-in-the-headlights look he wore during Beast. His hair has grown out, his face is not as drawn as it was over the summer. The slight adolescent acne, exacerbated by the sweat and grime and camouflage paint, is also gone.

The plebe has a bottle of Windex in one hand, a rag in the other: He sprays the sink and wipes it out, paying particular attention to the chrome faucet. His roommate moves even more quickly, using the flat of his hand to wipe invisible dust off the desk, off the tops of the computers. They straighten the shoes lined up beneath the bed, pull the blankets tight, make sure the towels in the towel rack are folded precisely in half. In between ministrations, they study their computer screens, where they've pulled up the front page of the *Washington Post*. They check their watches every few seconds.

In a rifle rack by the door, three M-14 rifles are locked with a steel bar and heavy padlock. The back of the door is decorated with an Army Football poster and a card with the heading *Risky Business*. "The decision to drink is risky business," the card says. "Leaders must assess risks and take appropriate action."

Like other colleges, West Point struggles with alcohol use by students. Because the environment is so tightly controlled, the opportunities to drink are not as numerous. But many cadets, when they finally get a chance to cut loose, do so with a vengeance.

"When most people think of West Point, they figure it's very conservative," Friesema says. "They think: no alcohol, no . . . immoral activity. But stuff like that goes on. It's bothersome, but it's not like I can go around telling people not to act like that."

Although he is far from the buttoned-up stereotype of the religious right, Friesema is a self-described conservative and a member of the Officers Christian Fellowship. He spends Sunday mornings teaching Sunday School to the three- and four-year-old children of faculty members.

"Let's go," Friesema says to his roommate. He has "FCDT," or Fourth Class Development Time. This is the time when team leaders and squad leaders quiz and instruct their plebes. FCDT is strictly limited to avoid abuses.

"Cadet time is the coin of the realm around here," a colonel in the Superintendent's office said. "The most precious commodity. If we add something to the schedule, something else has to go, so we're always careful about limiting things that might distract from the time they have to do their duties."

For the plebes, this means that the upper class cannot interfere with their study time. But cadets are still caught in a tug-of-war between the Commandant (military duties) and the Dean (academic duties). Cadets are always trying to strike a balance.

Friesema stands in front of his roommate, who checks his uniform. Good crease in the trousers, belt buckle shined and exactly aligned with the shirt placket, name-tag parallel to the shirt-pocket seam, shoes mirror-bright. Thus prepared, he steps out into the hallway, which is decorated with spirit posters for the upcoming Army-Air Force football game. All around, plebes greet sleepy upper class with, "Beat Air Force, sir!"

Team leader Huston, a yearling, checks Friesema's uniform, then asks about the continuing struggle between the president and the GOP. (It would be several months before Monica Lewinsky became a household name.) They talk about dirty politics and scandal mongering, about what the president should do. Every few questions, Huston tests Friesema's command of the military knowledge he's supposed to be memorizing.

"What's the armament on the main battle tank?"

The plebes are required to speak in complete sentences, without using abbreviations or acronyms, so the answers sound stilted.

"Sir, the main armament on the mike-one, alpha-one main battle tank is the one-hundred-and-twenty-millimeter cannon."

Huston talks about the different characteristics of smoothbore guns versus cannon. Then he asks Friesema to describe the other armaments on the tank, down to the nomenclature and ranges of the hand-held weapons carried by the crew members. Just your standard dorm hallway chat between a college sophomore and freshman. But the little tableau is indicative of how the old plebe system has given way to one that is more like the army.

During my plebe year, we had to report to the rooms of the yearlings, but the upperclass cadets often did not even get out of bed. We'd let ourselves in quietly and stand at attention in the darkness, backs pressed up against the wall lockers. Then a voice would come from beneath the covers, "Start the Days, beanhead." It was not at all unusual to recite an entire day's worth of plebe knowledge without ever seeing a face. On other days, we would prepare for the inspection, studying and memorizing, only to be chased from the room before we had a chance to recite it. This demeaned the plebes and devalued what we were required to learn. How important could it be if the inspector didn't even wake up?

The system still requires plebes to master the information. It also requires that the upperclass cadets learn how to conduct an inspection. After they finish talking politics and current events, Huston tells Friesema to recite "The Corps," which is a kind of second alma mater.

The company hallways are lined with posters and a few bulletin boards. One poster shows a close-up of a soldier; he is dirty, and his eyes, under the brim of his helmet, have that "thousand-yard stare" of the exhausted.

"Do your job!" the poster commands. "His fate depends on your skill!"

Huston gives Friesema a heads-up on what they'll talk about the next morning. Lately they've talked about articles from the front pages of the on-line newspaper. Over the past weeks Friesema has also had to learn about the infantry, armor, Special Forces, background information for next summer's introduction to the various branches of the Army.

After development time, Friesema continues the rituals of getting ready for his day. As he cleans he talks about the highlight of his semester thus far: his visit home over the recent Columbus Day weekend. "The people home are so much different from the people here," Friesema says. At first he thinks it might be the people there who have changed, then says, "It's probably me who's changed the most."

"I don't sweat the small stuff anymore," he says. "It used to be in high school if I got more than one project at a time I'd freak out. Here, getting more than one project at a time is the norm." He thinks it's ironic that he's been desensitized to small stuff precisely because his life is dominated by so many small demands.

Friesema checks a clipboard by the door, where an inspection log details what the inspector found wrong with the room the previous day. Cadet rooms are inspected daily by someone in the chain of command. He looks up from the notes, then tightens the beds and refolds a blanket at the foot of one of the bunks.

Friesema's bookshelf holds the one photo frame he's allowed. In it are smaller photos: He and a date dressed for the prom. He is thinner in the photo, and wears his tuxedo awkwardly. He and the girl stand before a small tree with a few tiny buds. Another photo shows two little boys, cousins who live on a family farm where Friesema hunts and fishes.

It was his experience as an outdoorsman that made him think he'd enjoy the field exercises at West Point. "I worked at a landscaping company in high school. My boss and his dad used to lease five hundred acres up in northern Wisconsin. They had this old school bus on the property. They took the seats out and put bunks in. We'd go up with a whole bunch of guys to hunt and fish."

He stands at his desk, his long body bent over, one hand on the mouse, scrolling through the on-line newspaper in search of some other bit of knowledge that will help him at formation, when he's expected to speak intelligently about current events. Through the window behind him, the sky begins to lighten. Taylor Hall, another castle look-alike, takes shape out of the gloom.

Before they leave for breakfast formation, Friesema and his roommate make one more sweep, blowing dust off desks, picking lint off blankets, making minute adjustments to the shoes and boots lined up under the bed. His roommate tosses a nylon bag full of hockey gear into an overhead locker.

"That's the only place they don't look," he explains.

Out in the hallway, an orange sign advertises a study session for an upcoming test in a plebe core course. In cheery script at the top, it says, "You Won't Have To Cram!" The company has an academic sergeant and an academic officer, junior and senior cadets, who help set up these kinds of sessions. They assign tutors and keep up with cadets in academic straits. In the hallway, a plebe in full uniform stands at each intersection. These are the "minute callers," human alarm clocks.

"Sir! There are two minutes until assembly for breakfast formation!"

They chant in perfect unison, in exaggerated slowness that sounds, at this hour, like some weird religious ritual.

"The uniform is: as-for-class under gray jackets! Two minutes, sir!"

Outside, under a clearing sky, the cadets gather in their hundreds for breakfast formations. Each company is assigned a specific spot on the paved space. The sergeants account for everyone, then come announcements about company activities ("There will be a 'spirit' push-up contest, Beat Air Force!") and the march into the Mess Hall.

The Cadet Mess is enormous: six cavernous wings, each three stories tall. Five wings are filled with ten-person tables where the four thousand cadets eat family-style meals. Most of the tables are organized by company, though varsity teams and some clubs have their own tables. It is a great coup for a plebe to land on a team table. Plebes can eat "at ease" with their teammates, and there is none of the performing that marks a plebe's life on company tables.

Friesema's table is in a wing dominated by an enormous mural, a sixty-by-thirty-foot painting of great military leaders from world history. The painting, which includes a cast stretching from Hannibal to Alexander, was a WPA project in the 1930s. Another wing that also dates from the twenties is dominated by a large stained-glass window showing George Washington in blue uniform. Wooden wainscoting lines the walls to a height of ten feet. Above that hang portraits of former Superintendents, the paintings shielded by Plexiglas since a food-fight some years earlier. Higher still, above the portraits, hang the flags of the fifty states. The space is as dramatic as any European castle; it is also, like any castle, dark and cold.

The plebes remove their jackets and drape them on the chairs. Hats go on a small shelf below. Friesema drops his gear and hurries to a table in one of the wide aisles to retrieve an armful of yogurt cups. Another plebe goes for coffee, but comes back empty-handed. In a cost-cutting move, there are fewer pots of coffee than there are tables; on some mornings they go quickly. The table holds cold cereal and packets of instant oatmeal. There are cartons of milk, a steel pitcher of juice, and a pitcher of water. In the dim light the tray of scrambled eggs brought by the waiter has a surreal color, like a pile of food dye.

The corps is called to attention by the brigade adjutant, whose command "Take seats!" is followed by the scraping of four thousand chairs on the stone floor.

The government provides $5.25 per day per cadet for rations. It takes some imaginative planning to feed four thousand–plus teenagers on a five-and-quarter a day. Since 1990, the Cadet Mess has also followed the Surgeon General's guidelines on fat and calories (less than 30 percent of calories from fat). Within these parameters, the Mess delivers between 3,200 and 3,500 calories per day to the average cadet, a little more in the summer when they are more physically active.

Attendance at breakfast and lunch is mandatory during the week; dinner is mandatory on Thursday night only. Cadets can attend optional meals (which are free) or buy from one of the many restaurants in the area that are happy to deliver pizza, sandwiches, or Chinese food. Visiting graduates notice that the meals have changed. Gone are the halcyon days of eating fat-laden foods and desserts, the years before every grade-school student in America started tossing around words like "cholesterol" and "arteriosclerosis."

The senior who sits at the head of the table is the table commandant. The table next to Friesema's is the "All Star Table." The cadet company commander sits at the head. The plebes sitting at the other end are there because they're the "problem children," the ones who need extra attention. At this meal two first class cadets are quizzing the plebe sitting at the foot of the table, who has been skipping class. Attending class is a military duty; cutting class not an option.

"He just wants to get thrown out," a second class cadet says as he spoons eggs onto his plate. "He should just resign."

The platters of food make their way to the end of the table, though no one is eating yet. When everyone has been served, one of the plebes holds the platter above his shoulder and announces, "Sir, there are two and a butt servings of eggs remaining on the table."

He puts the eggs down, and his classmate at the foot of the table says, "Sir, the fourth class cadets at this table have completed their duties."

The table comm looks up and says, "Eat."

The Mess Hall is a more civil place than it was twenty years ago, when maniacal upperclass cadets might spend an entire meal yelling at plebes for infractions real and imagined. Despite the complaints of old grads about the end of discipline, the Cadet Mess hasn't turned into Pizza Hut. When an upperclass cadet addresses a plebe, the plebe must put down his or her fork and look up. No eating until the upperclass cadet signals the conversation is over. Plebes no longer sit at rigid attention—which made it difficult to swallow—but they eat in silence, heads down and eyes on their plates. The table manners of the upperclass cadets vary from table to table; the standard is set by the table comm.

A few minutes into the meal and the cadet adjutant is back on the public address system. "Attention to orders!"

The four thousand cadets immediately stop eating during announcements of upcoming sporting events, lectures, drill periods. When the adjutant announces an Army victory in any sport, every cadet makes a fist and strikes the tabletop. There is a thunderclap *Boom!*; flatware jumps to the floor all over the Mess.

At the end of the meal, the plebes at Friesema's table stand, put their jackets on, remove their hats from beneath their chairs and, in unison, shout "Beat Air Force, sir!" before they scuttle into the press of gray headed for the door.

Friesema moves like a broken field-runner, dodging waiters and greeting legions of upperclass cadets with, "Good morning, sir!" or "Beat Air Force, ma'am!" Plebes are immediately identifiable because of the way they move, and often because of what they're doing. These blue-collar workers of barracks life must deliver laundry and messages, clean common areas, and generally provide much of the muscle power that keeps things moving. They are the privates in this military organization. They also get to exercise some leadership, as they are put in charge of other plebes to perform certain duties.

"Each week one plebe in the platoon is cadet-in-charge of duties. My team leader gets in trouble if I don't get my duties done right," Friesema says.

This is also a change from a time when plebes were blamed for, or at least hazed for, any failure. Now the chain of command is held responsible. This is a move not just to make cadet life more like the Army; it is part of the training of the upperclass cadets. They are responsible for what their subordinates do or fail to do.

"When my roommate, the football player, gets uptight, he starts yelling at people," Friesema says.

"When it was my turn to make sure laundry was delivered [to upperclass rooms], I found out how to do it, had [the other plebes] come to my room, and told them where laundry goes and sent them off."

Major Rob Olson is fond of saying that leading peers is one of the toughest challenges, because there is no authority from position or rank. Plebes will work for a classmate because they know their time in charge is coming, and because they respect the person put in charge.

"My roommate had the other plebes getting up at zero five hundred when they had until 6:30 to do a few minutes' worth of duties," Friesema says, shaking his head in disbelief.

The long hallways in Thayer Hall are full of cadets wearing the class uniform of long-sleeved gray shirt, black tie, gray trousers with black stripe. Name-tags go on the right pocket. Most wear bulky sports watches.

The classrooms are nearly identical: nineteen or twenty desks arranged in a horseshoe, an instructor desk and lectern, a TV monitor. There are no windows; the four walls are lined with high, freshly washed blackboards. (Each room has a bucket of water and a sponge, which the instructors use to wash the boards after each lesson.)

When Lieutenant Colonel Joe Myers, Friesema's math "P" (for professor), enters the room, the cadet in the first desk calls the group to attention, salutes, and reports any absences. Class begins exactly on time; every class starts the same way. This class happens to be all men.

Myers, a fortyish man with a small smile, begins by encouraging the plebes to bring their parents to the department open house during upcoming Plebe Parent Weekend.

"Your folks want to see what it is you do with your time," he explains.

At the visitors' desk is a black binder with a copy of the course syllabus. Friesema is in an advanced math class. On the board in the center of the room is a single sentence in Myers's neat hand, "Bacteria reproduce at the rate of thirty per hundred per second."

"What are some other uses for a continuous function?" Myers asks.

One cadet suggests the compounding of money. Myers talks fast, and the cadets take notes; inside of five minutes the front board is covered with equations. Every few seconds the P looks out at the cadets. In the horseshoe, everyone has a front-row seat, and Myers can spot a lost look in an instant.

"Are there any questions on the homework problems?" he asks.

Several cadets raise their hands, and the class works through the problems together.

Math instruction at West Point still bears the marks of the system instituted by Superintendent Sylvanus Thayer in the early days of the nineteenth century. Thayer, the "father of the Military Academy," decreed that each cadet would be graded every day. A key element of this system was the recitation, in which a cadet would be called upon to stand and explain some problem and its solution to the class, according to a very specific choreography.

Under a system used until just a few years ago, math instructors would take the report, then ask for questions. If no cadet asked a question, the next command was, "Take boards."

This sent each cadet to a section of blackboard (they were numbered at the top). Using a long ruler, the cadet divided the board in half from top to bottom. Then, still using the straightedge, he or she drew two rectangles in the upper left-hand corner. Name in the top box, problem number in the bottom. Yellow chalk was for the work; white, green, and blue chalk for the drawings. The answer was to be underlined twice in red and labeled "ANS." Naturally, the boards were graded for neatness. At some point the instructor would command, "Cease work!" All chalk went down immediately. The cadets stood by the boards as the instructor looked for a likely candidate to recite. Some instructors just looked at the work; others would also inspect a cadet's uniform, shoes, haircut. This was not Columbia, after all.

"Mr. Jones."

The cadets not chosen sat down.

Stand at attention next to the board; begin by saying, "Sir, in this problem I was required to find the area under the curve defined by the equation . . ."

Pick up the pointer; talk the class through the problem. Sometimes a helpful classmate might say, "Sir, I believe Mr. Jones made an error in the fourth step of the equation . . ."

These people were not popular, except as targets in plebe boxing. They all went on to Stanford and Harvard Business School.

Some instructors were interested in teaching math. They would gently ask questions about the method, about the reasons for using a certain approach. Others believed they were preparing the cadet to stand in front of a general at some future date and brief the plan for the next D-Day invasion. In that case, the cadet had better be ready to defend his work.

Murphy's law dictates that a cadet would be called upon to recite on exactly those problems she understood least.

"And why did you choose that method?"

Because it was my best shot, sir. The least complicated, the calculus equivalent of counting on my fingers and toes.

The system made cadets work, if only to avoid being humiliated in class. But it had its critics, too. West Pointers' reputation in the Army—they always looked for a definite answer, underlined twice in red—traces its roots back to that math board.

✥   ✥   ✥

After class, Friesema heads back outside. Everywhere, it seems, there are cadets running, singly and in groups. This is a constant at West Point, the sine qua non of cadet life. Some of them wear shorts and T-shirts; others wear sweats against the chill. They wear reflective bands like bandolleers across their shoulders. There is tremendous pressure on the women to be good runners, thus women are over-represented among the cadets who pass.

Friesema is excited about Plebe Parent Weekend, just a few days away. His parents are driving fourteen hours from their home in Racine, Wisconsin, to visit. They are, he says, very proud of his choice.

"It'll be nice to show them around here, give them an idea of how I live, what I do."

He is only a few months removed from the high school senior they knew, but he has a keen sense of how much he has changed. For one thing, being away from home has made him more aware of how much family means to him.

"My older brother, Andy, goes to the University of Wisconsin. We shared a room, and when he left I used to lie there and look at his empty bed. I really missed him."

It was Andy who got a mailer from West Point; and although his father tried to talk to Andy about the Academy, the card went in the trash. Bob Friesema fished it out and sent it in.

"My parents have my two younger brothers in private school, so my getting a scholarship helps. But it was harder for them to send me here than to send my brother to Wisconsin. R-Day was the first time I ever saw my dad cry."

Friesema, the second of four boys, is fairly self-aware for a teenager.

"There's a three-year difference between me and the next youngest, Dan. In the months before R-Day [the younger ones] really looked up to me, asked me a lot about hunting and fishing. I haven't really gotten along with John, the youngest, well. But in the short summer before I came here things got better."

Every few steps he greets an upperclass cadet with, "Beat Air Force, sir!"

"I picked on him a lot," he admits. "He was overweight. I was pretty ruthless. I feel bad about it now. He bawled his eyes out on R-Day, and he writes me e-mail more than anyone else."

The upcoming weekend will be a time to visit with his family. It will also mean a much-anticipated chance to leave the post. Although plebes do have some free time, they are limited in what they can do with that time, especially first semester. They cannot: wander into town, drive to a movie, wear civilian clothes, listen to music in their rooms, go to nearby New York City, visit friends at other colleges, have friends visit them overnight.

They can: play sports and work out. And study.

"I don't like the fact that there's no social life," Friesema says. "Zero."

If the lifestyle at West Point is dramatically different from his home life, the values are not. Friesema, a graduate of Racine Christian School, is completely at ease with the honor code, because it's in keeping with the values he was raised with.

"Some people don't take it too seriously, they think they can get away with little white lies. They don't always think that there are implications to what they're doing. Sometimes, with my roommates, I point out that there's a connection between their doing something and the honor code, and they're surprised. It just doesn't always occur to them."

In English class that afternoon, Major Julie Wright takes the report from the section marcher and has the cadets open their copies of *Newsweek* magazine. An article talks about the rise in co-habitation as an alternative to, or precursor to marriage. There are a variety of reasons for this change, the writer claims: decreasing stigma for couples who have children without marrying, rising divorce rates, decline of marriage as an ultimate goal, women's increasing financial independence, men's increasing independence from the idea that they need a housewife to take care of the home.

Wright asks the cadets to respond in their journals. Then she sends them to the boards to summarize their responses in groups of two or three. Bob Friesema's group writes on the board, "Marriage is a holy union of 2 people and should not be desensitized [sic]; it is necessary for family stability; unstable families lead to societal problems such as crime and are detrimental to children's development; cohabitation without marriage is the result of breakdown of morals and religion."

Other responses around the room are just as conservative. Wright does not comment on the content, but focuses on the rhetoric and the structure. Later, Wright says that the reaction was "not an anomaly."

"I definitely think they are more conservative than some of the enlisted soldiers they will encounter in the Army."

This is not surprising, given that the admission standards reward young men and women who have played by the rules, who have succeeded by all the conventional measures in academics, sports, and leadership. But the difference does contribute to the problems many young West Point graduates have communicating with soldiers and NCOs.

The long days of classes lead directly to athletics. Cadets who don't play on a varsity team (called Corps Squad) or on a club sport must participate in the extensive intramural program. During parade season (spring and fall), intramurals take place every other day; the alternate afternoons are spent at drill and ceremony practice. On any afternoon, thousands of cadets are out playing sports. Colonel Maureen LeBouef, head of the department of physical education (which oversees the intramural program), says that West Point has more sports opportunities for its four thousand students than Ohio State has for its forty thousand students.

This period of athletics extends into the evening, and cadets are still on the parade field as dusk gathers. A steady wind comes off the river and pushes yellow leaves into the artificial light from a practice field. There is a diamond-chip moon overhead, and clouds sail down the river valley. Under the lights of their practice field, the Army rugby team churns the turf into mud. Wearing throwback uniforms of striped jerseys and thick, knee-high socks, they slam into each other at full speed. When they pause, they breathe like horses after a gallop.

On the Plain, there is a heartbeat of a drum as plebes drill for Plebe Parent Weekend, which officially begins tomorrow, Friday. The windows of the Supe's house are warmly lit. A van drops off some MPs and a bugler for the simple ceremony that ends each day: The bugler plays "To the Colors," the retreat gun cracks over the valley, the flag comes down. People stop their cars and get out to stand at attention. The rugby team stops and faces the flag, and for an uncharacteristic moment, West Point pauses.

Just after breakfast on Friday morning, Jacque Messel heads off to the gym, carrying her athletic uniform in a mesh laundry bag. Like Bob Friesema, she is more at ease than she was in Beast, and has lost the fidgeting nervousness that followed her during the summer. The change is profound.

"The lowest point was my birthday [her nineteenth], August 17th. We were just starting the academic year, so I had all these new classes, and I had been through Re-orgy Week, and my parents had just left after their first visit.

"But it was also a big breakthrough, too. My Dad and I finally got to talk about my coming here, about my staying here. They said they would support whatever decision I made. I knew that all along, of course, but the visit just reinforced it."

She had been focused too much, she says, on Beast Barracks. Like many cadets, she had a hard time imagining West Point would be any different during the academic year.

"Platoon Sergeant Stitt helped me out there," she says, mentioning the universally respected first detail platoon sergeant. "He got out a copy of *Bugle Notes* and started looking at all the clubs and activities; then he asked me what kinds of things I'd like to do if I was still around West Point in the fall. He didn't try to convince me to stay; he didn't talk about my dad or anything. But he did help me see that there was life beyond Beast Barracks."

Messel hurries into Arvin Gymnasium. Parts of this old building resemble a cathedral or a monastery: hallways topped with barrel-vaulted ceilings, doorways guarded by bas-relief carvings of athletes. On a high wall near one entry is MacArthur's decree about the connection between athleticism and soldiering.

Upon the fields of friendly strife
Are sown the seeds that
Upon other fields, on other days
Will bear the fruits of victory

Messel heads to the second floor and Hayes Gym, a huge room on the second floor, where her class will practice one of the most grueling physical tests West Point offers: the Indoor Obstacle Course Test, or IOCT. Messel and the other women line up behind the men at the beginning of the course. As each pair of runners takes off (there are two parallel tracks), the cadets in line shout encouragement. Soon the huge old gymnasium is filled with shouting and grunting as cadets negotiate the obstacles.

Messel stands in line, bouncing up and down on the balls of her feet with nervous energy. At the "Go" signal, she throws herself down on the wrestling mat in front of her and crawls for twenty feet beneath a wire mesh stretched a foot or two above the mat. She runs through parallel lines of tires, the old football practice drill. Then an "elephant vault" over a padded frame that stands a good four feet over the polished floor.

From there she sprints to an eight-foot-high horizontal shelf. She grabs the edge, hangs, throws one long leg up over the lip of the shelf. Women tend to have more problems with this obstacle, as they do not have the upper body strength the men do. Messel uses her height advantage; she kicks, pulls, swings her way onto the shelf.

The cadets climb from the shelf over a railing and onto the second-floor track, then swing through a series of bars until they can jump back down onto the gym floor. More running. They race to launch themselves through huge truck tires that hang like tire swings from the ceiling. (One instructor said that the largest cadets, especially the football players, have a hard time fitting through.) After the tire they make their way—"walking" on outstretched arms, legs bicycling in the air—along two sets of parallel bars placed end to end.

By this time even the best-conditioned cadets are winded; they suck in huge gulps of the old gym's dusty air. The long muscles of their arms and legs suddenly feel wrapped in lead.

Thirty feet to the next vault: an eight-foot vertical wall. The technique is to jump for the top and place one foot as high up as possible on the wall. The sound of cadets kicking and crashing into the plywood wall punctuates the entire test.

Messel's height gives her an advantage, but eight feet is still eight feet. She drags herself over the top, drops down behind. Tiring, she runs flat-footed, her athletic shoes slapping the wooden floor. She climbs a padded step and reaches for the first rungs of a thirty-foot-long horizontal ladder. She swings out, legs churning as she grabs each rung and makes her way through space again. (Some of the gymnasts, their arms like steel cables, blaze through barely touching each bar.) Messel drops off the last rung and stumbles to the bottom of a frayed white rope. The cadets must climb sixteen feet to another shelf above their heads.

Messel grabs the rope, pinches it between her feet as she has been taught, and inches her way upward. She is not the fastest, but she moves forward relentlessly; it is, as the sports announcers say, a game of inches.

Sixteen feet is marked with an orange band. Messel touches the band, then swings her legs over to the shelf, climbs the railing again to the track. There is a canvas cart filled with leather medicine balls. They are just large enough and, at about twelve pounds each, just heavy enough that even the biggest cadets have to use two hands. There is no way to run smoothly with the medicine ball; it's like running while carrying a case of beer.

One lap with the ball (twelve laps to a mile on this track), then trade the ball for a baton, then a last lap empty-handed. An instructor with a stopwatch calls out the times as cadets cross the finish line;

they report this time to a recorder on the first floor. They are on their honor to report the correct time.

As they wait in line to report their scores, they rest, hands on knees. There is little talking. Messel has only a few minutes to shower and change after class. Books under her arm, hair still wet from the shower, she hurries to Thayer Hall for psychology, another required course for plebes.

The classroom is crowded and hot, and when she sits down, it is the first time she has stopped moving all day. Within a few minutes, several of the plebes, including Messel, are having trouble staying awake. This is a common problem at West Point, especially for plebes. Cadets operate on little sleep; they stay up late and get up early. If they sit down and remain still for more than a few minutes, they will nod off. It isn't unusual to look into a classroom and notice one or two cadets standing in the back of the room, struggling to stay awake.

After psychology, Messel has a free period, so she heads to her room. Like a lot of plebes, she lives with two roommates in a room designed for only two people. There is barely enough room to pull the chairs out from the desks. On the bookshelf above her desk is a large frame filled with smaller photos: a group of girls making faces at the camera; a small photo of Messel with her parents. In the photo she looks mature in an evening gown, makeup, jewelry. At West Point, there is no opportunity to dress that way.

"I've just tried to fit in, be one of the guys, I guess. I don't spend a lot of time with makeup or any of that. Even when we're allowed to wear civilian clothes, a lot of the girls choose not to let their hair down or wear makeup or jewelry. Some of them do take advantage of any chance, though, like at football pep-rally dinners."

(At the mandatory Thursday night dinners during football season, cadets sometimes wear various costumes, which may include civilian clothes.)

"I feel more comfortable just fitting in. A lot of girls, it really bothers them that we have to dress like the guys and all. Every time they go out they buy something feminine to wear or to decorate their room."

"Decorate" may be too strong a word to describe the individuality allowed cadets. Beside the one framed picture per roommate, the only other thing that qualifies as a "decoration" are a couple of two-inch plastic pumpkins tucked away on a bookshelf.

Messel checks her watch repeatedly as the morning slips by. Plebe Parent Weekend begins at lunch, and that's when she's due to meet her parents. She puts on her hat and goes outside to wait by the big clock outside her barracks. Her parents arrive a moment later, as if on cue.

Jerri Messel, Jacque's mother, is petite and pretty, a quiet woman. Physically, Jacque Messel favors her father, who is tall and broad. She does not have his booming voice and tendency to dominate the conversation.

Messel's demeanor changes around her parents. As she leads them into the Mess Hall for the guest lunch, she becomes quiet. Throughout the meal, the two women sit quietly as they hear stories of Robert Messel's cadet days. He talks about West Point in the 1960s, about playing Army basketball, about the great football teams, about the new buildings and the old way of doing things, about everything except his daughter's experiences in her new world.

At times, Messel seems confused about the role her father played in getting her to West Point. At times, she says he "encouraged" her, telling her stories about his experiences, about people he met and places he'd been. At other times, it seems that she became a cadet to please him. "This was his dream," she states flatly.

CRUCIBLE

The football season goes a long way toward making West Point a tolerable place to be stuck on fall weekends. Each week there is the buildup to Saturday, the pageantry of the pre-game parade, and the carnival atmosphere of the game itself. All cadets must attend, and tradition dictates that they stand for the entire game. The cadet section of the home stands is a solid block of noisy gray, and cheering for the team is a release for pent-up energies.

In many ways, the entire season is just a prelude to the Army-Navy game, which is played on the first Saturday in December. This final contest is almost a separate season. An Army team can go into this game with a losing record and still salvage a great deal by beating the midshipmen. (The Army-Navy rivalry is so central to sports at West Point that academy officials talk about the success of the athletic program in terms of percentage of wins over Annapolis.)

If football is a metaphor for war, then the Army-Navy game is the climactic battle. For the entire week leading up to the game, the corps wears battle dress uniform, part of the pep-rally feeling of one of the biggest weeks of the year at West Point. The barracks are decorated with huge bed-sheet posters hanging from windows (inside the quads; nothing is visible on the parade field side); plebes sound off with, "Beat Navy, sir!" at every turn. There are nighttime "spirit" missions and push-up contests and floats being made for the game. The Supe's house displays an eight-foot "GO ARMY" banner on one side of the porch, a "BEAT NAVY" banner on the other. The school colors of black, gold, and gray are everywhere. Formations are energized; plebes break ranks and climb on statues and balconies to lead cheers. All the months of hard work, all the routine and regulation, give way to college pranks. Bedlam reigns.

Except at the football tables.

While the corps is outside celebrating, the team is already eating in a corner of the Mess Hall wing dominated by the Washington stained-glass window. The players are excused from lunch formation so they can eat quickly and squeeze in some team meetings before afternoon classes. Their tables are piled with double portions: huge platters of gray meat, loaves of white bread in plastic wrap, lukewarm corn piled in rectangular steel dishes. The football players eat head-down; there is little conversation.

There is a notion popular among cadets who do not play intercollegiate sports that the athletes have life easier. No parades, fewer formations, all those road trips. Plebes on the football tables can relax, while a few tables away their classmates on company tables sit up straight, do table duties, eat in silence, and answer questions thrown at them by upperclass cadets.

The athletes are quick to point out that they undergo a grueling practice schedule. They return to the barracks exhausted and begin their studies late. There is no such thing as leisure time during the season.

Second class cadet Grady Jett, the Beast squad leader who stressed teamwork above all, sits at a mostly empty table in the stained-glass wing of the Mess Hall. His short hair is parted in the middle. The gold chain he wears around his neck looks out of place against his GI

T-shirt. All around him, plebes move cartons of milk and platters of food. Jett is disappointed with the team's record so far, but mentions his only touchdown, which came against Louisville the previous last game.

Football dominates his life. It is not the only thing he is concerned about, but it takes the most energy. And it is through football that Jett wound up at West Point. A successful high school player in football-crazy Texas, Jett worked hard to get a scholarship to ease the financial burden on his parents. He didn't know anything about West Point until Academy recruiters contacted him.

"At first, I was like, 'No way.' Then I started looking into it, into the possibilities for the future . . . it seemed like a good fit for me."

The biggest challenge for Jett was leaving his family; he is very close to his parents and sister. Amazingly, Jett's father attended every Army game of the season, flying to New York from Texas for the home games, flying around the country for Army's away games. The whole family will join him in Philadelphia after the Navy game.

But Jett has a few obstacles to clear before the weekend of celebrating. "This time of year the screws are really on," he says. He isn't talking about game pressure; he's talking about academics, about the projects and papers due near the end of the term.

Jett graduated eleventh in his high school class of six hundred, but at West Point he hovers around the midpoint of his class of just over nine hundred. He knows his grades could be better if he had more time to study.

"If I didn't have football, my grades would definitely be better, but I wouldn't trade it for anything. And it's not just football, but the friends I made, the experiences I had; highs and lows, working with a great bunch of guys."

Jett loves playing, but the time commitment isolates him from the rest of the corps. It is especially apparent to him this semester. In spite of the time drain of football, he was made a squad leader in his company. He is concerned that he cannot give the cadets in his squad enough attention.

"One of my guys had an older brother, a firstie, who died recently.

He'd gone on sick call and they told him he was dehydrated. They gave him some fluids and sent him back. His roommate tried to wake him up in the morning, but he was dead. I tried to help my guy get through that. But it's hard because I'm not around all that much. I try to stop by in the evening and see how the plebes are doing, but upper class aren't even supposed to be around the plebes during study barracks."

Jett eats quickly, then climbs six flights of stairs to one of the huge drafting rooms above the Mess Hall. Coach Darrel Hazzell smiles as he greets the players, then begins the meeting by reading a moment-by-moment breakdown of the upcoming afternoon practice. He hands out a photocopied sheet of neatly printed formations and pass patterns, then begins a rapid-fire patter that is almost completely jargon.

"Most of their guys are wrist-banded. In this non-crackable position," Hazzell says, pointing to the paper in his hand, "if his feet are inside it doesn't matter what alignment, you guys got to get inside. So make the switch signal. If we're in spread alignment, the back is not going to engage." The partial illustrations on the sheet have names like, "Rhonda 168 Hammer Smoke," and "Larry Special Left Pass."

They've been in the room less than ten minutes when Hazzell, still talking continuously, puts a tape in the VCR. The screen shows a practice. Hazzell is all energy, pacing and pointing. The cadets stay awake but have few questions; every once in a while they nod in agreement or understanding. Twelve minutes later the video is off and the cadets hurry out of the room, and on to their next requirement. As Jett enters the stairwell, the cattle-drive sound of the corps leaving the Mess Hall fills the building.

"Coach Hazzell is great," Jett says. "He's the best coach I've ever had. He never talks about personal stuff during practice. It's all football. The coach we had two years ago used to fall asleep during meetings."

From Washington Hall Jett makes his way to MacArthur Barracks and the dayroom of Company C4, which has been his company only since the end of August. In between their second and third years, cadets are "shuffled," assigned to new companies. The administration found that when cadets stayed in the same company for four years, as was the practice in the past, the companies developed "personalities." They could turn into subcultures that might or might not share the institution's values. Some companies were known for athletics or academic prowess; some were known for being hard on plebes, others for being easy on the fourth class.

From the administration's point of view, this meant a cadet who spent four years in one company might have a dramatically different experience of West Point than a classmate in another company. What's more, some of these subcultures were at odds with administration policies and even Academy values. Now cadets get to know more of their classmates. More importantly, attitudes across the corps are evened out, and the developmental experience is more uniform.

The dayroom, a combination meeting room and television room, is a windowless basement room with a pool table, a television, and VCR. Thirty or forty government-issue padded chairs crowd the space. One wall is decorated with a giant pencil sketch of a cowboy, the company mascot. Jett falls into one of the chairs.

The room is hot and close and crowded with second class cadets, none of them enthusiastic about this mandatory meeting about alcohol and drug abuse. A firstie asks the platoon sergeants for a head count and gets an "all present." Less than a minute later a cadet walks in.

"Good report," someone complains.

A second-class named Walt Hollis, who is the company First Sergeant, begins. "Welcome to another class on alcohol. Here's the plan. I'm supposed to show you this video, which lasts about twenty minutes, and then we're supposed to discuss it. I have a list of questions and discussion points. But the VCR is broken, so I'll just tell you about the video."

He reads the lesson plan from a set of printed notes. In the video he isn't showing, a student goes to a frat party and gets drunk to the point of alcohol poisoning.

Hollis looks up and asks, "Why do people drink?" There are a couple of chuckles, some whispered comments. There is still something in the air of the sophomoric fascination with drinking.

"People drink to escape reality," a cadet in the front row says.

"To hook up," offers another.

"Because we think we're much cooler when we drink."

"When does a person's drinking become a problem?" Hollis asks.

"When there's not enough to go around," offers a cadet slouched in a chair.

Someone in the back of the room asks, "Why do you all have to act so stupid?"

"A lot of people say that West Pointers get into the Army and don't know how to act," Grady Jett says. "They have a drinking problem because they've never been allowed to do it normally; it's always such a big deal."

The juniors talk about a cadet who graduated a few years back. He used to return to the barracks drunk and fling open the plebes' doors and shout at them before stumbling off to bed. They generally agree that this person was a sad case and no kind of leader, but there are still a few lingering sniggers.

The juniors talk about the resources available on Army posts to help soldiers with these problems, then Hollis ends by quoting the lesson plan: "Alcohol and drug abuse are inconsistent with military service."

"Alcohol is a problem at West Point," Hollis offers. No one disagrees.

"There's binge drinking and underage drinking and DUI. How can we work against that?"

A young woman offers, "They're on the right track in giving cadets more freedom. People take responsibility for their own actions. It's not like before, where cadets got out and just exploded."

A firstie closes the class. He is soft-spoken, makes little eye contact.

"Have a good time this weekend [in Philadelphia at the Army-Navy game], but be smart about it. Think about what you're doing before you get into trouble."

Later that afternoon Jett heads to Michie Stadium for practice. He has spent three hours in class today. He will spend four and a half at the football complex.

The training room buzzes with dozens of conversations. Jett climbs on a table and shaves his recently injured ankle with an electric razor; a trainer will wrap it in a small cast made of white athletic tape. While he waits he removes his BDU blouse; he is not big for a Division I player, just a shade over six feet and 180 pounds. Army recruits from a small pool of talent: kids who can play Division I football and handle West Point's academics. They are the same athletes being recruited by Navy and Air Force, by Brown and Penn and Duke. Players recruited by the academies must also consider the service commitment of at least five years.

The cinder-block walls are decorated with paper signs, most of them martial.

"The more you sweat in peace, the less you bleed in war."

"Tomorrow's Battle is Won During Today's Practice."

A large plaque lists the years in which Army won the Commander-in-Chief's Trophy, awarded to the major service academy that defeats the other two in football. Army last won in 1996.

Jett is closer to the young men in this room than to anyone else at West Point, but he also talks with some envy of his friends who aren't on Corps Squad. "They have the time to go out with their buddies, to make other friends in the corps. I know my roommate, when he comes back from dinner, he has some time to hang out with other people and make other friends. I get back and have to get to work right away. I wish I had time to know more people."

The trainer works quickly, unrolling tape and tearing it into neat strips.

"People who know what we go through respect us. Then there are those people who think we just come up here and joke around and that we get out of a lot of stuff, like drill and formations, that we're just get-overs. They don't understand how hard we work."

"On the faculty you see both attitudes. Some of them know what we go through. Some of the Tacs went through our strength-training workout one day. The coach said a lot of them couldn't even finish it," he says with obvious relish. "That changed a lot of attitudes. A teacher traveled with us when we went to a hotel the night before a game; he was amazed at how intense our schedule is. We go from practice to meetings. He never knew that we went through all this prior to games."

Jett jogs off to change into his practice uniform, then moves to the warren of meeting rooms, which are shabby, with low ceilings, holes in the walls, burned-out lights. The floor is thick with folding chairs, and after the players start to assemble, with piles of shoulder pads and gold helmets. The linemen lumber through the room; a few of them have stomachs like middle-aged men and will have to lose weight to fit within the Army height and weight standards before they can be commissioned. There are more motivational posters hung unevenly around the tight space. One could be Coach Hazzell's motto.

"Success is not an Accident; It's a Planned Event."

Hazzell coached at the University of Western Michigan, where he "had a lot more time," he says. "It took a while to adjust to the cadets' schedule. We coaches have to understand what these guys go through, because their lives off the field are so hectic, there are a lot of demands. But they're outstanding guys to work with. They have this tremendous willingness . . . they want to work, on a consistent basis. A lot of places you gotta pull teeth to get things going, but not here. You can't ask for a better coaching situation."

Hazzell is five ten, with a neatly trimmed mustache and a black Army Football nylon warm-up suit, a baseball cap with a simple "A" on the front. He begins the session with his minutely detailed overview of the upcoming practice session, then he goes to the white board and sketches the plays they will work on today. He talks fast, quizzing the players and shuffling through a pile of papers on a rolling table.

Grady Jett, who was up until midnight or 1:00 the night before, who was up this morning at six, and who just returned from a Thanksgiving leave where he got "no sleep," is fighting to keep his eyes open. Yet he answers every prompt.

At 4:20 P.M. they are on the field for warm-up exercises. The practice field sits in the right angle formed by Michie Stadium and the Holleder Center. (Sports venues are named for cadet athletes later killed in action. Dennis Michie, USMA 1892, brought the new sport of football to West Point in 1890. Michie was killed fighting the Spanish in Cuba in July 1898. Don Holleder, USMA 1956 and Army quarterback, made the cover of *Sports Illustrated* magazine during his last season. Holleder was killed in action in Vietnam in 1967. He was thirty-three years old.) The big lights come on and the temperature starts to drop; when the players stand on the sidelines and remove their helmets, steam lifts off them in clouds. Beyond the bare trees the mountains to the east are outlined against a dove-gray sky.

The practice field is two hundred yards long and seventy-five wide; it is covered with players, some in white practice jerseys, most in black. There are more than a hundred and eighty cadets in this system, more plebes than upper class. This week's practices are emotionally charged: because it's the Navy game, and because it's the end of football for some of these players.

"Seniors," one of the coaches says, "This is your last Monday practice."

The entire two-hour practice is run like the two-minute drill at the end of a clutch game. The coaches read from identical copies of the time plan. They move squads and subsquads here and there on the field at a run, bringing together the backs, then the receivers and quarterbacks, then the linemen.

"Everything is designed to get the most out of the time we have with them," Head Coach Bob Sutton says, "because their schedule is so tight."

And it works. The parts all come together like rapidly spinning gears, like a complex field operation, like a battle drill. This is due, in part, to meticulous preparation. But none of this would be possible without the tremendous resources available to this program. There are two elevated platforms, raised by hydraulics, at one end of the field. Three video operators spend the entire second half of practice shooting footage from fifty feet up. There is a squad of managers in yellow sweats, a dozen trainers and coaches.

Football gets the most attention at West Point—many feel too much attention. It also generates revenue that supports the rest of the varsity sports program. Football is the national spectacle, the perfect mix of violence and glory and sex for our society. Maybe it works at West Point because it is a military metaphor, a surrogate for war. Or it may be that football, in spite of commercialization, retains some of the magic Cadet Michie saw when he organized the first games on West Point's muddy Plain.

Late in practice, Grady Jett catches a touchdown pass, the ball floating into his hands in the artificial light, in the crisp air. It's a Hollywood play, of course. The defenders aren't running full-speed and the point is execution, but the play is beautiful nonetheless, and the players on the sidelines cheer wildly. This is Navy week.

After practice, the players shower and gather in the big hallways of the Holleder Center, where long tables have been set with dinner (the rest of the corps has already finished eating). The meal set for them is huge: plate-bending servings of chicken tacos. As with the Mess Hall meals, there is a scarcity of fresh vegetables. The closest thing here is shredded iceberg lettuce, which is nearly white under the fluorescent lights. Jett eats sparingly for someone so active.

"A lot of people who know what we go through up here respect us for it. Up here," he says, nodding at the practice field beyond the dark windows, "we've worked together through all those practices: when it's hot, when it's freezing cold. Teamwork is about pushing other people to get better for the sake of the team, about pushing yourself to make yourself better individually. . . . That's the kind of leader I want to be, not just bossy, but someone who earns respect. Not by being their best friend, but by being approachable."

Jett is slowing down. His day started early and he hasn't stopped moving—rapidly—since.

"I've got a group coming to my room at 8:00," he says, looking at his watch. It is 7:30. "We have a fluids design problem due this week.

We have to come up with a way to check the viscosity of an unknown fluid."

He'll be up until midnight, or later. And at 6:30 tomorrow morning, the plebe minute callers will be outside his room, signaling another day.

Like Grady Jett, Shannon Stein, the second-detail Beast squad leader, is also a varsity athlete. Recruited to play soccer, she is the workhorse of the team in the contest against Navy just days before the Army-Navy football game. Stein, who is five three and weighs a hundred pounds, plays as if she weighed two hundred. She is a fierce competitor, throwing herself at the ball, at the opposing players, rolling and tumbling, everywhere at once.

But the effort is not enough, and Army loses. Stein, who wore her emotions on her sleeve during Beast, is subdued the next day. She wears her camouflage BDU pulled low enough to hide her eyes. With ten unstructured minutes before lunch formation, she heads to MacArthur Barracks and the three-person room she shares with Meghann Sullivan and Lynn Haseman.

The shades are pulled down and the room dark as Stein enters. Haseman is stretched out on one bunk, fully clothed and asleep. Sullivan, sitting in the glow from her computer screen, is dressed in BDU pants, olive drab T-shirt, combat boots. She is broad-shouldered, an athlete, with a fine complexion and a Botticelli face. The paper she is writing for PL300, the standard leadership course for second class cadets, is due at 4:00. In trying to explain what it's about, she stumbles through jargon about "AOIs" and leadership theories and external outputs of something called "OOSM", then she switches to plain English.

"It's about how West Point turns out so many social retards," she says.

The downside of living in an isolated, insulated world is that many cadets learn to function only in that small universe.

"You have some weird people here," Sullivan says. "You have a lot of people who think 'I'm this great athlete,' or 'I'm really smart', but a lot of them have absolutely zero social skills. They think everything revolves around the rules and that the rules are an answer to everything."

There are rules governing nearly every aspect of cadet life, from the way they dress to how they keep their rooms to how they address each other on official business. But it is not just the presence of so many rules that makes cadets feel squeezed; it is the lack of privileges. Only first class cadets are allowed to leave post daily, and then only after their last duty. Evenings and weekends away from West Point are meted out to the under three classes, with the plebes having the fewest opportunities to escape. A cadet who is failing a course or is under some disciplinary action loses most of these privileges.

"You're twenty years old and you have to get permission to go out at night," Sullivan says. "You should be responsible enough by this time to say 'Well, I have to study tonight,' or 'I can afford to go out.' They baby-sit you."

Lynn Haseman stirs under her comforter, then sits up on the bed and looks around. Like Sullivan, she wears BDU pants and an olive T-shirt; her boots are beside the bed.

Haseman is from an Army family. A third-generation West Pointer, she is the ninth person in her family—and the first woman—to wear cadet gray. Her grandfather is buried in West Point's cemetery, her cousin is a tactical officer. Her father and General Christman, the Superintendent, were on the track team together as cadets. Her mother and Sue Christman, the Supe's wife, were sorority sisters. Her father was the former Commandant's squad leader; the current Commandant is a classmate of her uncle's. In Shannon Stein's description, she is "Miss West Point."

"She knows *everybody*," Stein says.

"I still have to do everything everyone else does," Haseman says. She is matter-of-fact about it, not at all defensive.

"The only way it would help her is if she was about to get thrown out," Sullivan says.

"I'd call everybody I knew if they were trying to throw me out of here," Haseman affirms.

The women are not speaking hypothetically. Sullivan and Haseman are both facing disciplinary action for fraternization.

"I just got into trouble for something so stupid," Haseman explains. "I went and bummed a cigarette off a plebe . . . this is like three months ago. And two or three months later a classmate of mine who doesn't have the intestinal fortitude to come up to me in person writes a 'serious observation report' on me because he found some rule that he thinks I broke."

Sullivan, who only stood in the doorway of the plebe room, is not in as much trouble as Haseman. Both are angry with the classmate who reported them because he did it in a sneaky way: he did not confront them, or even ask them what happened. He merely turned them in.

This is a problem with an institution that relies so heavily on regulations: it can create martinets who look to the rule book for the solutions to any problem. A mature "leader of character" would have had the guts to confront the rule-breakers.

"This might be the biggest leadership school, and these guys are going to go out there and be CEOs, but they don't have the people skills to handle tough situations," Haseman says. "They think the answer is writing a report on someone. This guy didn't even *ask* me about what happened."

She stands, tucks her T-shirt into her pants, pushes loose ends of her hair behind her ears. She is lively and pretty, in a tomboy way, with light brown hair, fair skin, and the build of a serious athlete.

Shannon Stein, still leaning on the windowsill, is quiet. Meghann Sullivan is animated; Lynn Haseman is angry. This is not just about oppressive rules; it's also about gender, about how men and women treat each other, about how well they follow the strict rules while living in such an intimate setting. Cadet women are on the front lines of the battle to integrate the Army.

"If a girl does one thing, she's labeled," Sullivan says.

"If I walk to class with a guy, then I must be dating him, I must be having sex with him," Haseman says sarcastically. "One of the biggest problems in civilian life, in business, is the integration of women. Here we are, twenty years after the first women came through this place, and we're still dealing with the same problems. It's kind of ridiculous that we have to put up with the same stuff."

"I had a teacher who just didn't like girls," Sullivan says. "There were two of us in the class and he called us stupid right to our faces. The guys just laughed about it. No one wants to see what really goes on."

Haseman folds her comforter neatly and places it on the foot of the bed, grabs her boots and sits on another bunk to pull them on. "Whenever I get stressed out I go and sit in the bleachers, across from the barracks, and I look back at this place and try to be objective about it. I mean, it's very pretty here, and my parents and my grandparents got married here."

She is now in a complete uniform: hair tucked neatly into a clasp on the back of her head, boots shined, hat in hand. "If you don't want to quit at some point," she says philosophically, "there's probably something wrong with you."

"You need to take a break every once in a while," Sullivan agrees. The alternative is the West Point cocoon, the "social retards."

"You should see these guys," Haseman says. "All they think about is the Army. They don't know who they are so they let West Point develop them, develop their personality, and it shouldn't be that way. When that happens, you wind up a complete dork."

Lunch formation is on the concrete apron between MacArthur Barracks and the Plain. Stein stands in the rear of the company and watches as twenty or thirty plebes, shouting "Beat Navy," rush the half dozen seniors who make up the battalion staff. They tackle the first class cadets, one of whom is a tiny woman who puts up a great fight. The plebes sing Army fight songs. One climbs the base of Washington's statue and leads a cheer.

Inside the Mess Hall, Stein makes her way to the soccer team's table.

"I could get a lot better grades if I wasn't on Corps Squad," she says as she waits to take her seat. The women's soccer team played eighteen games between August and November. Practice for the next season will run through the spring.

"Our coach was paranoid after last season." (The team finished 2-16.) "He just keeps us running the whole practice: sprint drills. My legs are killing me." There are several nods around the table.

Stein's boyfriend, another second class cadet, comes to the table, whispers something to her, stands around for an awkward moment, then leaves. There are pitfalls to dating cadets, Stein says after he's gone.

"Everybody knows your business. And some people who are dating, they just spend all their time with each other. Then if you break up, everyone knows all about it."

But with almost nine men for every woman, it is almost impossible for a woman to escape attention.

After the meal, another male cadet comes to the women's soccer table. He stands behind the chair of another player, looking out of place. She doesn't look up. When the seat next to her opens up, he sits down and starts talking.

Though the Mess Hall is huge, and is packed with four thousand young people all dressed alike, it is still possible to run into people. Stein sees plebes from her Beast squad from time to time. Often they greet her with their squad motto, "Always the Hard Way!"

"I was amazed at how flexible I could be," Stein says of her summer experience. "They [the new cadets] could be scared of me—I could spit fire or be relaxed and still keep tight control. I had one plebe [to supervise] last year [before her assignment to the Beast cadre]. But to take a group and have them do things together and have them enjoy it, that's a great thing. I cared about them; they learned their stuff well. People in the chain of command got all wrapped up in 'finish the mission, finish the task;' it was more important that [the new cadets] learn."

"I'll probably be a platoon sergeant next semester. I want to get up in front of a group of people before I go out on CTLT. [Cadet Troop Leader Training sends cadets into the Army to fill lieutenants' jobs for a month.] I'd rather mess up in front of plebes than in front of privates who will judge West Pointers by what I do."

Stein joins the camouflage throngs squeezing through the big doors as the Mess Hall empties.

"I'm not afraid to fail at stuff. I know it won't be the last time. I took a lot away from that experience at airborne school [in which she had to repeat a week's training]. I wasn't leaving there without my wings, even if it took all summer."

She checks her watch. At West Point, it's always time to hurry somewhere.

"So many people here are afraid to fail. They've always been the 'good cadet.' But it makes a person humble, and you need to experience that before you go out into the Army. Because the people I'm leading will have had some failures and they need to be able to talk about it with someone."

Kevin Bradley, the firstie who commanded Alpha Company for the first detail of Beast, is now the executive officer, or XO, of the second regiment, a five-stripe captain, second in command of a thousand or so cadets. Moving that many cadets to Philadelphia for the game is a military operation that involves scores of buses, dozens of memos, hundreds of e-mail orders, coordination meetings and checkpoints and emergency procedures. In addition, second regiment has "the duty" this week, which means all the cadets on guard—and there are quite a few—come from Kevin's unit. They will run a command post at the stadium, they will be ushers and courtesy patrol (think unarmed police enforcing good manners); others will return to West Point immediately after the game to oversee those cadets on restriction, who do not get a weekend pass and must return to the Academy right after the game. Although he works through an extensive chain of command, keeping this many moving parts going eats up a great deal of time.

Still, Bradley seems to be enjoying himself as he stands in the sunshine for another lunch formation. Part of this is due to his competence: he gets the hard things done and makes them look easy.

"Kevin is one of those cadets who get it," Major Rob Olson says of him. "Things come to Kevin. Or at least, things appear to come to Kevin."

Bradley watches a thousand cadets of second regiment form into eight companies, leaving an open arena in the middle of the paved expanse. Everyone is in camouflage battle dress uniform—part of the "prepare for combat" mind-set that reigns this week.

More spirit activities. The company commanders wear the costumes of the unit mascots: one is wrapped in a robe and carries a spear, like a Roman foot soldier. Another is dressed as a dog, still another as a bear. In the ranks there is an excess of good cheer; the weather helps: It feels more like October than December.

When Bradley hears the story of Bob Friesema's English class and their surprisingly conservative views on cohabitation, he isn't surprised.

"When you're a plebe you don't want to deviate. A lot of what they say is what they think the P [professor] wants to hear. Then in philosophy [a required course for yearlings] they pry away and get you to ask tough questions. If you had a room full of firsties, you'd get a wide range of answers."

Even so, Bradley agrees with Friesema's English professor that the cadets are more conservative than the soldiers they'll lead. Kevin doesn't have a lot of time with Regular Army troops, but he finds hope in his limited experience during summer training. He could relate well to the junior enlisted soldiers, mostly because he is close to their age. The difficulty came in working with the non-commissioned officers, the sergeants who are, for the most part, older than the lieutenants.

It is not uncommon for a new lieutenant, at twenty-one or twenty-two, to be paired with a platoon sergeant in his early thirties. The sergeant has all the experience, and a smart lieutenant knows that. But the officer is in charge. Clever young officers figure out how to walk the line—deferring to experience, soliciting input from the old hands—without turning over command. Good sergeants know how to train their lieutenants, passing along knowledge, providing guidance and coaching, without creating a power struggle in the unit. This working relationship is critical to how the Army's small units function, and the West Point's administration knows that. In the last few years, the academy has brought more NCOs on board to help cadets get it right.

"We have a little spirit thing here at lunch," Bradley says. "A pugil stick fight between a female midshipman, one of the exchange people, and a woman cadet."

The academies exchange first-semester juniors each year; West Point hosts six Air Force Academy cadets, six midshipmen, and two cadets from Coast Guard.

A cadet in a referee's shirt steps into the center of the paved arena and introduces the women. The midshipman (the Naval Academy doesn't call them midshipwomen) is much taller than the cadet. The pugil sticks are about the length of an ice hockey stick, with heavily padded ends and more padding wrapped around the middle. The contestants wear thick gloves, like those worn by hockey goalies, and helmets with wire face masks. Rock music blasts from a portable speaker system as the two women swing the sticks inexpertly, thumping each other with the padded ends.

Major Rob Olson joins Bradley. "We had to go with pugil sticks because the jello didn't set in time for wrestling," he says.

When the midshipman lands a solid blow, the crowd cheers her. The cadet doesn't give up, and keeps wading in for more. After three short rounds of mostly ineffectual pushing, the referee declares the midshipman the winner. There are a few jeers, but the regiment applauds. Bradley ducks into the Mess Hall just ahead of the wave of camouflage, and Kevin goes back to talking about his summer training, when he spent time with a U.S. Army detachment in Japan.

"It would have been nice to work more with the junior NCOs," he says. "I mostly dealt with officers. My big concern is this: Here [at West Point] we get all the technical skills, but we don't have any real experience dealing with soldiers' problems. That will be the hardest thing for us."

Over lunch—small pieces of beef floating in gray-brown gravy—Bradley reflects on his experience working for Major Rob Olson.

"He let us take charge of things, and he made it so you didn't want to let him down. For instance, he gave me the plan from the previous year's R-Day. He said, 'See what's good, what's bad. Even though the old one was good, I'm sure you can come up with something better. You can change it if you want; just be able to explain why.'

"Sometimes Tacs get too involved. Major O walks this line," Bradley says, making a knife-edge with his hand. "Letting us do it on our own and letting us screw up. I tried to give the platoon leaders the same sort of room he gave me. And then, if one of them had a problem and went to the Tac, Major Olson would look at them and ask, 'What did Bradley say?' "

The opportunities for leader development are limited during the academic year, when the Dean takes over and the focus of cadet life is study. But the Commandant's side looks for ways to continue military training. From the cadet perspective, this means that someone in power is always adding to the list of requirements.

"They put in something called unit training time," Bradley says, "on those days when you're not playing intramurals and you don't have drill. Some companies do road marches. We did classes on the M9 [the Army-issue pistol], PT. Those are times I get to do what an XO does in the Army, oversee what the staff is doing to make things easier for the companies. The company training officers write plans for the training. It's an exercise in leadership for the upper class, and it's a good review for the plebes. It's all the stuff they're going to use [in summer training]."

The cadets don't have a problem switching from calculus to field-stripping a weapon and back to American history. But they hardly welcome the additional work.

"Everyone was cynical when they introduced unit training. That used to be free time you could use to work out, visit friends, play a sport for fun, read a book, sleep. I know they're trying to walk a line between providing training opportunities for cadets and overloading them."

This is one of the things that differentiates Bradley from younger cadets or classmates who haven't had his responsibilities. Having been in charge, he knows there is another side to the story, and he tries to understand both.

"Trying to make things work is a lot harder than just standing in ranks bitching about everything," he says.

Despite the uniforms, formations, and afternoon training, there is a letdown in military posture that accompanies the academic year. "The third class team leaders are supposed to pick up [training the plebes] where the Beast squad leaders left off. But it takes a lot of creativity to keep it challenging."

In years past, the upperclass cadets' job was simply to make plebes suffer. Now they have to set and achieve goals; they have to plot ways to develop—West Point's favorite word—their subordinates. Leaders are given limited time and resources, and must design training that will instruct and motivate. This is exactly what NCOs and junior officers do every day in the Army. West Point cadets are saddled with the task at the beginning of sophomore year—one plebe at a time. Kevin Bradley didn't start out in this system, but he's bought into it.

"At the end of plebe year you're sick of getting yelled at. With our class the big thing was 'Do you have your socks pulled up?' All the plebes had to wear our athletic socks pulled all the way up, but the upper class didn't. It was a stupid rule, besides being unfair. You look on the surface of it, you just don't see any lesson there."

Now, when he watches a leader, when he looks at his own leadership abilities, he asks what the leader is doing to move people along, to make them better soldiers.

"If you just look at the surface of things, if you don't look for the 'why' behind stuff, it's easy to be cynical. It's also easy to be cynical if you're the kind of person who tends to blame the system for all your problems.

"And then people just get tired of all the rules. When you're a plebe you're just scared you're going to do something wrong. You don't expect to go out at night and you're too busy to worry about it much. By the time you're a yearling you've figured out how to live here and you get bored. You know what goes on at other schools, all your friends tell you constantly what you're missing. Then you start asking yourself 'Why can't I walk out into Highland Falls [outside the gate] at night if I want to?' "

Like a lot of cadets, Bradley worries that the sheltered cadet existence does some harm.

"Most of us are social idiots," he says. "We don't know how to interact. Everything is taken care of in our world. Like your money, that's a perfect example. You don't have to learn how to budget because it's all done automatically."

As with meals, barracks, laundry, medical and dental care.

"We go to parties at other schools and the West Pointer tries to make up for a few months of being locked away by drinking everything in sight. So the cadet passes out and misses the party and manages to look stupid in the bargain. The kids from civilian schools know how to handle themselves."

After lunch, Bradley walks the circuit around Trophy Point. The river is gunmetal blue where it splits the mountains to the north; on its surface, boats large and small cut white wakes. On Clinton Field cadets have piled brush and wooden pallets for a pep-rally bonfire. Topping the mound is the hulk of a wooden sailboat painted battleship gray.

The Navy game is another milestone in Bradley's last year as a cadet. His parents live only fifteen miles from Philadelphia's Veterans Stadium, the site of the game, so they've been hosting parties since Kevin's plebe year. This year's will be the biggest celebration.

"We're going to have nine people staying at my house over Army-Navy weekend. There'll be guys sleeping on the floor of every room. My mom's been cooking all week. She loves it when all my buddies come over."

Bradley can practically smell graduation week, though it is a long winter away. He has a perspective on his experience that the younger cadets don't have yet. "In the beginning all your friends at civilian schools are telling you how much fun they're having, and you're just sitting around here. Going to the gym is considered a good time. But now, four years later, they're still doing the same things, and I've had opportunities that they haven't had. Like the chance to be a company commander this summer, and the chance to go to Japan for five weeks. And I know I'll have a job when I graduate."

The firsties have already chosen their branches. On his uniform Bradley wears the crossed cavalry sabers and superimposed tank of the Armor branch. He will be a tanker, and he intends to go to Europe. Cadets choose their first assignments based on class rank. Bradley, who ranks high in his class, is confident he'll get the assignment he wants. He is looking for adventure, for a challenge.

"When I said I was going to Germany, my folks were like . . ."

He makes a face; this couldn't have been the best news for his parents. "But then my mom said, 'Hey, I've always wanted to visit Europe.' "

"I'll probably go to Bosnia, and that will be good experience for a lieutenant."

At the beginning of December, Alisha Bryan is only days away from her twentieth birthday; yet she carries herself with the confidence of a woman ten years older. Her background also sets her apart. Most cadets come from white, middle-class suburbia. Bryan is the daughter of a mixed marriage, a black father and white mother, who spent her high school years in a mostly black high school in inner-city Atlanta.

Bryan, who was Alpha Company's counselor (the milk and cookies lady) during Beast, is five foot seven, with dark curly hair and caramel skin. On the bookshelf of her room in Scott Barracks, which is bright with morning sunlight, is a large frame with a photo collage of Bryan and her friends. One high school photo shows her as a teen, with long hair falling in ringlets beside her face. In another photo she and her mother are clasped in a tight embrace.

"My Mom comes to West Point and meets all my guy friends,"

Bryan says. "She just loves hugging them all, then she'll say to me, 'Alisha, you should have about ten boyfriends here.' "

Joking aside, the respect she has for her mother is clear in every sentence. Bryan's mother set an example of bravery for her. Until she was fourteen, the two lived in Alaska with Alisha's father.

"Life with Dad in Anchorage wasn't great," Bryan says as she checks her uniform in the mirror before heading out to formation. "He was a pool player; basically, a high class bum. He never let me do anything. My freshman year of high school, I wasn't allowed any phone calls, no dances, no school activities."

Alisha rankled under the strict controls, but they decided to leave when he became abusive. "We took off in the middle of the night, the day after my fourteenth birthday. We went to Atlanta because the city is more tolerant of multiracial families, and because my godmother lives in the South. My mom's mission was to set me up, give me whatever advantage she could. She made sure I could fend for myself."

Outside the open door to her room, plebes thunder down the steps to formation. Their combat boots rattle the metal stairs.

"Beat Navy, ma'am!" they say as they pass Bryan.

Although there were some financial struggles, Bryan relished the freedom of her new life. The deal her mother made with her was simple: Alisha could participate in as many activities as she wanted, as long as she kept her grades up. In what turned out to be excellent preparation for the pace of cadet life, Bryan held down several jobs and ran on the cross-country team.

Because she is articulate and attractive, and because she is a success story, West Point gave her an extra couple of days' leave before Thanksgiving and sent her home to inner-city Atlanta to talk to high school kids about West Point.

"It was kind of weird going back. The last time I was there I was mostly worried about how to get out of class and hang out at the Taco Bell."

If she cut classes, it didn't hurt her performance: she had a 4.0 average and good SAT scores. As a black woman with good grades, she had a wide choice of schools to choose from. But West Point's concept of all-around development appealed to her, and the Academy was the only school she applied to.

"My mom got [me] into the University of Georgia," she says with a smile. "She filled out the forms and everything. She was worried that I needed a backup.

"If I'd have gone to a regular school it would have been work and study and that's about it. I wanted something a little different. I wanted to see if there was something out there to push me."

Bryan sounds a theme common among cadets: *I came here because I wanted something different; I wanted a challenge.*

As she steps out into the sunshine, North Area is already filled with camouflage uniforms.

"I came up here for a visit and thought it was the most beautiful place I'd ever seen. It was a nice spring day, warm like this. I had friends here who showed me around. And there were old grads in the Atlanta area who contacted me; I was impressed by that. They cared about their school that much to go out and try to get good people. Not all schools did that."

The pool of qualified minority candidates is small. West Point admissions has several officers who concentrate on nothing but recruiting those men and women. But they are competing with every other top institution in the nation that wants a diverse student body.

At the beginning of her Beast, there were only fourteen black females in Bryan's class; this was quite a change from her high school, which was mostly black students. Still, being part of a double minority didn't present any problems for her. There was some extra attention she didn't need, but that was all flirting.

"When I was a plebe the women had to live near the first sergeant and CO [cadet company commander]. They acted a little like big brothers," she says, smiling. "They were always checking out who was coming to visit, making sure that no one was bothering us."

Bryan's independent streak did not serve her well as a plebe.

"The upper class had this idea of what the ideal West Point plebe should be; I didn't gibe with that ideal. When I was a plebe I was the one with an attitude because I didn't act the way others did. I was used to standing up for myself, which meant I was the one who would speak up if I thought things were wrong. I knew the limits of what the upper class were allowed to do, and if they pushed it or crossed the line, I'd say so. It definitely made life harder for me."

Bryan stands behind her platoon, her eyes gliding over the ranks, checking haircuts and uniforms. She is still checking out the troops when she says, "Race issues are resolved more easily here than gender issues."

"Some people think women don't belong here. That's hard to change. My Sosh P [Social Sciences professor] talked about how it used to be in his company, which he called 'Boys One.' They had segregated hallways; no women in some places. They told the women to 'get rid of the wiggle.' "

She laughs at the story. She's talking about boys' school pranks, nothing to be taken too seriously. Finally, it all comes down to one measure. A woman who can cut it physically is accepted. Bryan, a runner in high school, was better prepared than some. She fits in well enough that she sometimes doesn't even notice right away that she is the only woman in a class.

"I guess I felt it more plebe year. Like, in class, if we were talking about rap music, the P would say, 'Cadet Bryan, what do you think?' Like, 'Let's get the black woman's perspective.' "

She laughs, then gets serious.

"I haven't felt any discrimination here at West Point."

Two drummers beat a cadence, which echoes off the stone walls as the companies of fourth regiment march to the big Mess Hall doors. The plebes walk like robots, head and eyes rigidly to the front. During lunch Bryan talks about her involvement with the Contemporary Affairs Seminar, a black cadets' group. The group does community outreach programs and hosts a conference in the spring to which they invite inner-city high school students for discussions about college and other opportunities. The conference serves a purpose for the cadets, too, Bryan says, "So we don't forget where we came from."

"We're ground-breakers. It's good to remember how far we've

come. I mean, I talk about having *only* fourteen black women in my class—well—Henry O. Flipper was the only black cadet at West Point."

Henry O. Flipper, born in slavery, endured racism and a lonely existence to become West Point's first black graduate in 1877. Bryan and her friends know that they are carrying on in the tradition of Flipper, of the drive for civil rights. She is stirred by this being part of something forward-looking. In her world view, the future is better than the past.

On the way out of the Mess Hall after lunch she passes a firstie, and Bryan says hello. When he is out of hearing she points out that this black cadet is the deputy brigade commander.

"I heard there was even a black first captain once," she says.

Vince Brooks, '80, was the first black cadet to wear the six stripes and gold star of the first captain. Brooks, who became the youngest colonel in his class, later had lunch with the Contemporary Affairs Seminar. He told them, bluntly, that they had to be able to handle the visibility that comes with being a minority.

"It's an extra responsibility," he told them. "So what? Ruck up [strap on your rucksack, i.e., shoulder the burden], or find something else to do."

Outside, Bryan joins the fast-moving throng headed for the academic buildings. She talks about race as unself-consciously as she might discuss the weather. But not everyone at West Point is so relaxed. Bryan heard one white cadet comment, "You see a bunch of white guys sitting at a table after dinner, you think: lacrosse team. See a bunch of black guys sitting around a table after dinner, you think: coup."

"That's probably true," she says. "But it's really pretty simple. You want to talk to people with similar backgrounds, similar interests. It's like New York City with its Italian neighborhoods, or like when an American overseas wants to talk to other Americans, just for something familiar."

Bryan is consistently cheerful, and, like Kevin Bradley, is able to put her cadet experience in a larger context. Rather than complain about her lack of freedom, she chooses to focus on the advantages of living in this tight community.

"My friends at regular colleges have a hard time with money. They have to work at a job or two, go to school and study. They have it harder than I do. At West Point, if you just follow what they set out for you, you don't have to worry too much. I don't have to worry about eating Cup-O-Soup for days at a time."

Her friendships are a big part of what she values about West Point. She knows she has more men friends than she would someplace else, but that doesn't mean she's ready to date cadets. Many of those couples wind up at the altar immediately after graduation, something that Bryan pronounces "stupid."

"I'll only be twenty-one. What if I want to go to Germany or take advantage of other opportunities? A lot of people are in a rush to get married. It's convenient, because you're both here, you have so much in common already. But then you head off to different assignments and you say, 'What was I thinking?' "

"Once I have a family that'll be my main priority. The army is what I chose for this part of my life; I'll only be twenty-six when my commitment is up. There's lots of time—if I decide to get out—to do whatever I want."

"I think it's important to have a positive attitude, to be pleasant. It's still going to be stupid stuff whether you're bitching about it or not. Why punish everyone around you who has to listen to you?"

Bryan tends to look at things from this human relations point of view; she is glad she spent her summer counseling new cadets. For her, West Point is about the lives she touches and the lives that touch hers.

"This place is about service. I wanted something more than a regular college experience. I wanted to learn about myself and be challenged. I've come to like the leadership aspect of this place; it suits me," she says. "Like being platoon sergeant. I get to decide how I want to do things, how to run inspections. The platoon leader gives me autonomy. And when he asks a question and I can say, 'Already thought of that; got it covered.' That's really cool."

THE FRONT RANK

Cadets call the period between Christmas leave and spring leave "the gloom period." Everything at West Point is gray: the uniforms, the buildings, the sky, the somber mood. Unlike the fall, there are no home football games, with their invasions of visitors. There is just the relentless cold: cold floors in the barracks, cold wind blowing through the big doors of the Mess Hall so that the cadets nearest the entrance huddle in their coats until long after the meal has started. There is a long stretch of work before Spring Break.

On this bitter morning in January the temperature hovers at about twenty degrees, though the wind rocketing down the river valley makes it feel worse. It snowed most of the night, three or four inches on top of a layer of ice left over from a previous storm. A few lonely cars sit in the parking lots, and most of the civilian workers have been told to stay home.

But the cadets are out. In their woodland camouflage BDUs, with their shoulders hunched up against the weather, from a distance they look like a herd of two-footed animals looking for cover. They stream out of the barracks by the hundreds, by the thousands. Most of them head for Thayer Hall, the four-story former riding hall turned academic building that sits perched on the edge of the flat ground. Inside, the fourth class cadets of Company F-2—The Zoo—prepare a defensive plan to close the border of the fictitious republic of "Magapa" from a hostile mounted force.

A young woman, seven months out of high school, stands in front of the classroom, her map overlay projected on the big screen beside her. A river cuts across the small-scale map from northwest to southeast; a small town lies at the center where a bridge, carrying a north-south road, crosses the river. Low hills squat beside the blue water where she has placed her infantry squads to control the river crossing. She gestures at the map, unsure of herself.

"The La Costan forces," she says, naming the fictitious enemy, "probably won't use the road." She points to another route from the north. "They'll . . . uhm . . . come this way?"

It comes out like a question. The other plebes in the room, who are also completely new to this, offer no encouragement.

With her finger she traces a route that will keep the approaching enemy hidden behind a ridgeline until they are up to the bridge, almost directly across from the American position where this platoon leader-in-training has positioned her meager force.

"We've . . . uhm . . . we've plotted artillery fire back there," she says, "because we won't be able to see them if they move up that way?"

There are small crosses on the map overlay marked "TRP," for target reference point, pre-planned targets for the artillery. The platoon leader on the hill south of the river might not be able to see the enemy, but with a good plot and a radio, she'd be able to rain artillery fire on anybody stirring around back there.

The plan is a good one. It lets the Americans control the river crossing without sitting right on top of it; they won't get pinned down in the low ground around the bridge. But the delivery isn't inspiring.

Sergeant First Class Jonathan Brown, the Tac NCO for F-2, tells her so.

"What's with that timid little voice you're using?" he asks.

"I have a cold."

"Too bad. You're out in the field; I'm out in the field. You're cold and wet; guess what? So am I."

Brown, who has been a field soldier for most of his fifteen years in the army, sits in a chair in the middle of the room. He looks around the room, where the cadet desks are arranged in fours and fives. This is a teaching point about conditions in the field, not an attack on a sniffly cadet.

"You've got to put that information out there," he says. "You've got three squad leaders, three type-A personalities; all of them want to get out on that hill and start getting ready. Tell them what they need to know, then let them do their jobs, right?"

The young woman nods. She is already moving back to her seat, ready to be out of the spotlight. The next plebe who briefs, using the same slice of map and same scenario, tries to sound more confident. In addition to the squad positions, he has also indicated where the platoon's critical anti-tank weapons will be positioned.

"You probably don't want to stick your CP right on top of that hill," Captain Brian Turner, F-2's Tac, says from the doorway. "Why is that?"

The briefer turns to the screen. Sure enough, he has put the little symbol for the platoon command post smack on top of the high ground.

"They'll see you from a mile away," another cadet says.

"Right," Turner says.

Turner, who was the associate Tac of Alpha Company during Beast Barracks, is a tanker. He knows how tank commanders scan the terrain in front of them, looking for the obvious: a lookout on the high ground, spotters near the hilltops.

"What else?"

"Artillery?" a cadet attempts.

Turner nods encouragement. The cadet goes on.

"They're going to plot artillery on the hilltops, too," the plebe finishes. "Just like we do."

For almost fifty years, U.S. Army training exercises were plotted on maps of Europe; everyone knew who was coming over the hill. The generic replacement for the Soviet army—the motorized rifle and infantry regiments of La Costa—is a notional enemy. Today's plebe tacticians could call it a "fill-in-the-blank enemy": Somali warlords, Serbian police forces, Haitian mutineers.

The cadets do not ask about the scenario; after all, this isn't a class on world politics. This preparing to fight fill-in-the-blank enemies will create an Army that can deploy anywhere and, on command, shoot up whoever happens to be coming down the road to the river.

"Firehose learning," Turner says during a break. Cadets pass in the hallway. The background music is shuffling boots, winter-cold sneezes and coughs. This is military intersession, a two-week mini-semester during which cadets take only classes in military subjects.

"The team leaders shoot a lot at them [the plebes]. They've come a long way in a short amount of time as far as their knowledge goes."

"Team leaders work with the plebes on the capabilities of a specific branch," Sergeant First Class Brown adds. They cover a different one each month. For instance, when they were working on their knowledge of the infantry, they had to learn all the weapons, their ranges, what they could do. Here, we're starting to put all that knowledge together."

Brown is a big man with coffee-colored skin and large hands. He looks like he could be Eddie Murphy's big brother. "This summer, they'll get out on the ground at Camp Buckner and see that there's a lot more to it than drawing some symbols on a map. But this is a good start."

Turner and Brown are in the front rank when it comes to both teaching cadets the skills they'll need, and to setting the example of how officers and non-commissioned officers should act. Turner works with the first class cadets, who are closest to becoming lieutenants. He also spends time with the plebes, so he can get a sense of what the company looks like from the bottom up. Brown, the "real" non-commissioned officer, works with the yearlings [sophomores] and cows [juniors], who hold NCO rank in the corps.

Like most of the officers at West Point, Turner believes that the strong presence of senior NCOs like Brown has added immeasurably to the cadets' preparation. They leave the Academy with a clear understanding of how lieutenants and NCOs should work together. But working with senior NCOs can only accomplish so much.

"Cadets are a little isolated," Brown says. "They don't know how to talk to privates, because they haven't been around privates. They've been around cadets; they just see officers and NCOs."

Graduates of the Army's ROTC program aren't as isolated. In fact, they often spend their summers and time out of class working at jobs—like flipping burgers—that put them right next to the kind of young man or young woman who comes into the Army as a private.

Turner walks out into the storm and heads back to Bradley Barracks, the same building Alpha Company occupied during Beast. The blowing snow is channeled by the buildings, and Turner pulls his hat low to cover his eyes and returns the salutes of cadets coming in the opposite direction. His greetings are informal: *How ya' doin? Hey, how are you? Hi, there.* He is the friendly lord of the manor.

The barracks are warm compared to the storm outside. They are also scrupulously clean. One of the two hallways in the company area is crowded: Old desks stand along the walls, replaced by new desks with computer platforms. Cardboard and pieces of packing crate are stacked haphazardly.

"All this stuff has to go," Turner says aloud to some cadets.

Nearly every doorway in the company is decorated with a photocopied flyer. At the top, it says "F2 Zoo" above a photo of an adult and baby gorilla. Beneath the photo, lest a visitor think this was a college dorm, is the company's mission: "To conduct operations that promote academic, military, physical and moral/ethical excellence to prepare its soldiers for future leadership roles in the Corps of Cadets and the United States Army."

Below that, the acronym "METL." In the Regular Army, METL

stands for "Mission Essential Task List," those things the unit must be able to do to fulfill its wartime mission. The METL for F-2 reads:

- Foster Academic Excellence
- Prepare Soldiers for immediate and future military leadership
- Instill a competitive spirit within the Zoo that demands physical fitness and athletic prowess
- Reinforce a moral and ethical climate based on the Seven Army Values
- Protect the force

"The cadets came up with this," Turner says of the fliers. He steps up to one door and absentmindedly straightens a wrinkled corner of the paper. Turner got this process started and held the cadets responsible, but he doesn't take credit.

His office is at the end of the long hall. Inside, he flops down on one of the two vinyl, government-issue couches. Like most other Tacs, he's decorated in the "I-love-me" style: There's a guidon from the tank company Turner commanded, a group photo of his Ranger school class, a framed certificate of completion of the Special Forces Selection Course at Fort Bragg. The walls are a kind of equivalent of the badges soldiers wear on their uniforms: here's my military biography, my curriculum vitae.

"This company had a kind of bad track record for academics," Turner says. "I picked a good firstie to be the academic officer, and I told him: 'You'll get graded on your performance. Make it work, you'll get the A [in military aptitude].' He set up a system to monitor people who were having problems. Like in physics and chemistry, he identified those yearlings who were doing well. Then he took them up to the Center for Enhanced Performance."

The center is a laboratory, staffed by civilian educators, that uses the latest in psychology and pedagogy to help cadets boost performance in academics and even in sports. F-2's academic officer got all of his tutors trained there, then posted a schedule of the exams in major courses. He pushed the chain of command to get involved: Squad leaders and team leaders knew how their people were doing.

The year before Turner arrived, F-2 averaged thirteen or fourteen course failures a semester. In the semester just ended, there were seven. Turner is proud of the record and pleased that it was the cadet leadership that marshaled the company's talent. The chain of command even did the unpopular stuff, enforcing study conditions in the barracks.

"When some yearlings are goofing off in the hallway and making noise, [the leaders] say, 'You got something better to do?' We had all these people worried that grades would fall off, you know, since the plebes had TV cards in their computers, phones in their rooms. Looks like we're going to have forty-five or so on the Dean's List."

The door to the office is open; a cadet comes to the door, pauses.

"What's up?" Turner asks.

When Turner greets cadets he often slides into Army-speak: The aphorisms and colorful metaphors that pour in a solid stream from some people as soon as they put on camouflage. When Rob Olson— white-bread son of suburban Minneapolis—lapses into this, it is mostly the common Army-Southern hybrid: lots of twang. Turner's army-speak has a hint of black English. "What's up?" comes out close to "Waz-*up?*"

The first class cadet enters the office and hands Turner a three ring binder that says "Pass Book." This is how cadets request weekend passes and provide addresses and phone numbers of where they'll be while away from West Point. Some of the entries are a little sloppy.

"I don't want no Sanskrit in here," Turner says, pointing to an illegible entry. He quizzes the firstie on the approved passes. Did he check on eligibility? Is the system he used fair? A big part of his job, as Turner sees it, is to help the cadets connect what they're doing with what they'll do in the Army. This is also a way to fight cynicism among cadets.

"They complain about having to do some stuff. They say, 'How is this going to help me in the Army?' And I tell them that lieutenants have more than one job. You have your go-to-war stuff: are you technically and tactically proficient? Then you have all that other stuff. You've got to take care of soldiers, take care of families, keep track of equipment, help people plan their careers, all of the other stuff."

"Cadets respond to responsibility," he says. "I put one of my more cynical cadets in as company XO [executive officer, second-in-command]. People said, 'Whoa, you kiddin' me?' The guy's even on the overweight program and might get launched out of here. But he's stepping up to the plate."

Turner's job is to help all of his cadets learn how to succeed. His favorite tool is the one-on-one counseling session. He pulls open a drawer filled with neat folders, extracts one, and lets it fall open in his hand. It is a cadet performance record. Stapled to one side is a spreadsheet that shows the cadet's academic and military grades, summer assignments, academic major, hometown. There is a small black-and-white photo in one corner, a counseling form on the opposite side. Just above a job description that lays out what the cadet is supposed to do in support of the company mission, there are two blocks filled with handwritten comments. One is labeled "Strengths," the other is "Needs Improvement." The handwriting is the cadet's.

"I have them fill this out. Then I tell them what I see and we talk about what they need to do to improve. A lot of leaders . . . don't keep their subordinates informed. I guess people think of it as a confrontation. But if you let them know the score right at the beginning, if you take the time to do that right, when snafus come along you just go back to the original and say, 'This is what we agreed to, this is what you're showing me.' You can't be afraid to confront people. You have to have the ability to look someone in the eye and say, 'You're not cutting it. You're not making the standard and this is what you need to do to fix that.' "

Two cadets appear at the door, and Turner motions them in. First Class Murphy Caine, the cadet company commander, begins talking as soon as he steps in the office and, although he doesn't interrupt, rarely stops. The other cadet is a junior, Company First Sergeant Cedric Bray. The two cadets, who hold two of the most important leadership positions in the company, look like testimony to Turner's broad reach in looking for leaders. Caine is small, with dark hair and fair skin, energetic, constantly moving. Bray, tall and black, speaks deliberately and watches everything around him.

Caine walks over to Turner's chair behind the desk.

"Can I sit in the power seat, sir?" he asks. He sits and places the palms of his hands flat on the desktop. Turner asks about a meeting scheduled for that evening. Caine and the other upperclass cadets will brief the company on their plan for the semester: responsibilities and expectations for the chain of command and each class. They've culled the guidance given them by Turner and the higher echelons and have broken that down into a series of practical, usable steps.

"What's your plan?"

"I'm going to talk about the company mission," Caine says from behind Turner's desk. "Then the company staff is going to get up and talk about all the pieces, about how we're going to make that happen."

"Give me an example," Turner says.

"Well, the academic officer will talk about tutors, about how we're going to run that program. The platoon leaders will talk about conditions in the barracks, study barracks, and about passes. We'll show how we got input from the chain of command, how our goals fit into what came down from higher."

The cadets used the Army manual for laying out the plan. Turner is pleased because the cadets have accomplished two missions: they have created a plan for running the company for the semester, and they have learned how to plan using the Army model. It's not Desert Storm, but in eighteen months, some of these cadets wearing the black shields of the first class will be in Bosnia or the Sinai or wherever the current hot spot happens to be. They'll be ready to contribute.

Turner leads the two cadets into the hallway as they talk about the state of the company area. The Tac speaks to every cadet he passes; he knows names, hometowns, what subjects they're good in. In between chatting up passing cadets, Turner quizzes Caine and Bray on the new furniture: how many desks have come in? How many are still missing? Bray, a good First Sergeant, knows what's going on in his company and answers every question. Just as the cadets are learning some new skills, Turner has challenged himself as well. He is preparing to be a field grade officer (major and above), exerting his influence indirectly, leading through his subordinates and resisting the temptation to jump in and do everything himself.

Turner leads the cadets into one of the latrines that run down the center of the building. Each latrine is also a locker room, with several dozen gray metal lockers for athletic equipment, racks for drying wet clothing, showers.

"Why do you inspect?" he asks the cadets. "What are you looking for?"

"You want to make sure people have the equipment they need," Caine responds. "That it's in good shape, that they're taking care of it."

"Right. This isn't about 'You're a dirtbag.' " Turner says.

The lockers do not have doors. Gym shorts are on one side of the shelf, athletic T-shirts on the other, socks rolled into tight little balls. Athletic shoes go on top of the locker, aligned and facing forward. Taped to one end of a set of lockers is a photocopied page from cadet regulations: Appendix D, Annex B, United States Corps of Cadets Standard Operating Procedure. The sheet has a line drawing of an athletic locker. The shirts and shorts in the illustration are drawn with a ruler; the real lockers are almost as neat and are arranged in exactly the same way.

The barracks PA system announces, "Third regiment lunch formation goes indoors."

"Force protection," Caine offers. "No sense in standing around in the wind and freezing cold, maybe have somebody slip on the ice and get hurt."

The cadets don't remember having any formations moved indoors last year. The year before that, one or two were moved indoors. Some cadets see this as wimping out. After all, wars aren't cancelled because of weather. But leaders also must take care of soldiers and avoid unnecessary injuries. There is always a clash between the "drive on" attitude—we're tough, we don't give in, we don't give up—and common sense.

Back out in the hallway, the three men talk about bulletin boards. Caine is particularly proud of one labeled "community activities." A banner reads "Support Breast Cancer Research," just above a wrinkled piece of foil tacked to the center. It is the top of a yogurt container. The cadets learned that the yogurt company makes a donation for breast cancer research for every top sent in. Since the yogurt cups are served in the Mess Hall by the thousands, the cadets have made substantial donations.

Half of the bulletin boards are covered with candid photos of the cadets, along with printed biographies on three-by-five cards that state hometown, favorite sports, favorite quotation, branch the cadet wants to join. One cadet has put up a photo of a classic Ford Mustang; another has a close-up of himself at the helm of a sailboat. There are group shots: barracks birthday parties in which the celebrant is covered in shaving cream. One young woman has a picture of herself in an evening gown, long hair draped to her shoulders, Hollywood smile in place. Another board has two small snapshots. In each of them a plebe stands with his back to a wooden locker in a cadet room while two other cadets—upperclass, judging by their demeanor—stand on either side, one talking into each ear. The pose is a classic tableau of plebe year.

This is the posture plebes assumed when an upperclass cadet said "drive around to my room." The unlucky plebe stood up against the locker (or the wall) and steeled himself for what was coming, which could be a simple demand for recitation of fourth class knowledge to a screaming match between two cadets—one on either side—that was meant to rattle, demean, and sometimes reduce to tears the plebe caught in the crossfire. One graduate in his early forties remembers being invited by upperclassmen to "come hang around my room and listen to music." When he showed up, the cadets indicated the doors of the wardrobe, which opened out. The plebe draped one arm over the top of each swinging door and hung there—

with the sharp edges of the door biting into his armpits—while the upperclass cadets challenged him to see if he could last through an entire song.

But the upperclass cadets in this picture are smiling. It's a game, meant to be ironic ("this is how it used to be") and a threat ("and we could make it this way again"). The plebe, however, doesn't look amused. The whole point of being a plebe is powerlessness. Things happen *to* plebes. The plebe in the picture has no control over whether this game turns nasty. Turner taps the photo. "That's me," he says.

The plebe has a full head of hair; Brian Turner keeps his head shaved. The plebe in the picture is thin; the officer standing in the hallway now has spent a lot of time in the weight room. But mostly the plebe looks scared; the captain is the picture of confidence.

"I was playing a little game with the cadets. I didn't tell them much about myself. I sort of kept it a mystery, like the captain in *Saving Private Ryan*. And sure enough, they wanted to know more. Somehow they got my mom's phone number back in Chicago, called her up and got some stories. Next thing I know they're busting on me with dirt from when I was a kid. They got this," he says, indicating the bulletin board photo. "They got another picture of me and superimposed my head on Mr. T's body."

"You have to laugh," Turner says. He is smiling, but the amusement doesn't run deep. "They [the cadets] look to see if you're comfortable enough with yourself to have a sense of humor."

"I didn't have a lot of confidence in myself as a kid. That was one of the things I was looking for when I walked through the gate here."

The other thing the young Brian Turner was looking for at West Point was a way out of his Chicago neighborhood. "The guys I went to high school with were punks. I didn't know what I wanted to be, but I knew what I didn't want to be."

Turner bounced around among three different high schools as his mother, raising him alone, struggled to support Brian and his younger brother. Turner was always the new kid in school, and he stuck out because of his interest in reading and in getting good grades. Then a cadet came to visit Kenwood Academy, where Turner spent his last two years of high school. Impressed by the sharp cadet in the neat white-over-gray uniform, Turner started reading more about West Point and the Army.

One afternoon, he was working at his job as a ticket-taker at a local movie house when a man came in wearing a BDU field jacket. Turner saw the man's Ranger tab and commented on it. The impressed customer became even more animated when he heard Turner was interested in West Point. The movie patron was Kit Bonn, '78, who was then in graduate school at the University of Chicago. Bonn invited Turner to meet him for lunch on campus, and the two talked about West Point until Brian was even more convinced that he'd found his course.

"I didn't come to West Point thinking, 'Let's see how this goes,' " he says.

He chuckles at his own bravado as he remembers meeting his Beast roommate. "Our first day here, and I announce, 'I'm gonna graduate from here, and I'm gonna get my diploma from the President.' "

Predictably, his roommates laughed at him. They were focused on getting through the next day of Beast. Four years later, President George Bush was the guest of honor at Turner's graduation.

"Usually the speaker just gives out diplomas to the top guys," Turner says. "But that was an election year, and I guess he figured it was good campaigning, so he gave diplomas out to everyone. And my old roommate was like, 'You made that call four years ago, man!' "

The barracks are slowly filling up with cadets returning from class. Turner returns to his office, which has a view of the hillside behind the barracks: rocks, ice, steadily falling snow, and another brick wall. If Eisenhower and MacArthur Barracks have million-dollar river vistas, the view from the back side of Bradley Barracks is tenement airshaft.

Turner is single, the only single male Tac in the corps, living in an isolated community of mostly married people. He hates that part of the assignment here. "Professionally, this is a great assignment. I believe in this place and what we're doing here. Personally, it's tough. Almost all of the officers here are married. There's no O Club to speak of," he says, referring to West Point's Officers' Club. At other posts, particularly the large installations, the Officers' and NCOs' clubs act as social centers: low-cost country clubs with frequent dinners, dances, and even shows. West Point's club is open for lunch and the occasional dinner.

The end of the working day at Fort Hood might see the officers of a unit head off to the O Club bar for a beer. It is this team identity— I'm part of something bigger than just myself—is what many soldiers like best about the service. That spirit is harder to find at West Point, at least among the staff and faculty.

"Working with the cadets is great, and I try to stay busy that way. I help out with the marathon team, I'm helping set up Ring Weekend, the Martin Luther King Day dinner. I work with the Contemporary Affairs Seminar and I travel around with the admissions folks. Still, at the end of a long day it's pretty tough to go back to that empty apartment."

Single officers live in the Bachelor Officers' Quarters on South Post, on the grounds of a defunct women's college West Point acquired for its facilities. Apartments for majors run to about nine hundred square feet: a bedroom, a living room, a galley kitchen, and a tiny bath. Some of them have river views; half of them overlook the parking lot for the Visitors' Center.

Single officers are, in some ways, second-class citizens. For instance, Turner can host cadets at his apartment, but they must come in groups of two or more. Married officers don't need chaperones. The message the Academy sends with this policy, which is just knee-jerk overprotectiveness, doesn't sit well with Turner. The young men and women who aren't allowed to visit Brian Turner's quarters unescorted will spend their summer parachuting out of airplanes and helicopters, driving tanks, even walking patrols on Korea's demilitarized zone. He is entrusted with their development, but isn't trusted to be alone with them.

Turner might be expected to react to such a slight by withdrawing from the cadets. Instead, he becomes more involved. This week he is helping the Contemporary Affairs Seminar plan the dinner to honor Henry O. Flipper, West Point's first black graduate. All the Academy brass will attend, he says, but there is not much mixing of the races off duty. Turner even hears some complaining that black male cadets don't choose the Army's combat branches in any great numbers.

The combat arms—infantry, armor, field artillery, combat engineers—are the specialties most people think of when they think of soldiering. These are the branches whose job is to kill the enemy. They are also the largest branches, and the branches from which the Army draws most of its generals. But many cadets—black and white—don't see the combat arms as preparation for civilian careers. Turner believes that West Point contributes to this problem in its minority recruiting practices.

"First, any African American high school student who's good enough to come here is also being recruited by other colleges: Howard University and Brown and Yale. West Point tries to get them from the same angle: the marketability of a West Point education. We say, 'This is great preparation for life, for a great career. A free education at a name school.' So we wind up with kids who are already thinking about what they're going to do when they get past this little five-year commitment."

"I'm tough on African American cadets," Turner says. "I try to make it clear to them that they're always on display. There are three things people told me to be careful of, especially as a black officer.

"First: never be late. Second, you've got to write and speak properly."

Turner is "on" all the time. He naturally wonders if the first thing someone sees is a black officer, instead of a captain who happens to be black.

"And the third thing is: be a leader."

Yet Turner acknowledges that some of the cadets, seeing his success in a branch dominated by white officers, think of him as an "Oreo," a sellout. He doesn't spend a great deal of time worrying about it. He is more concerned that many senior black officers won't become involved in events like the Flipper Dinner because they're afraid people will think of them as militant.

Turner gets pressure from both sides. A white lieutenant colonel in the Tactical Officer Education Program told him he wears his race on his sleeve. Some black cadets think he should wear it more prominently. When a black woman cadet was accused of shoplifting while on leave (a violation of the honor code and grounds for dismissal), Turner was convinced she was guilty: her performance in school went down, and she withdrew from her friends. Although the honor board didn't find her guilty, the cadets knew where Turner stood. Other African American females thought he should have supported her based on race alone. To him, the assumption made by those young women—that he is an African American before he is an Army officer—is troubling.

"My concern for the Academy is that we *don't* think race is an issue anymore."

This isn't an intellectual exercise for Turner. It touches the way he interacts with cadets, the way he approaches his work.

"I get here early and I leave late," he says. "I have to work twice as hard to get the same respect, and it's tiring being in a fishbowl all the time. But that's OK; I can handle it. I've run into some things that were tougher to handle." In his first unit, Turner's immediate superior, his company commander, told him he was doing a great job as a lieutenant. But his Officer Evaluation Report was not stellar. The company commander told Turner that the commander at the next level (battalion) had made some comment about race. Turner approached the battalion commander, who was white, but the only feedback he got was, "You have an attitude problem. You walk around here like you're better than everyone else."

"I was pretty disappointed that that was the best feedback the system could give me. I mean, how are you supposed to work with that?"

Turner kept on working hard and was vindicated in his next report. More importantly, he made something positive out of the experience.

"That's why I spend so much time counseling cadets. That's why I don't hesitate to tell them exactly how they're doing, exactly where they stand. That's the only way they're going to be able to perform. And they've got to learn that doing that, telling people where they stand, is part of the leader's job."

It is time for lunch. Turner stands and goes through the small ritual of putting on his camouflage field jacket: First he tightens the drawstring that pulls the waist in tight; then the zipper, then the buttons, all the way up to the collar. He is not compulsive, but he is meticulous, and, like a good soldier, he wears his uniform smartly.

"I learn so much from these kids," he says, studying his black leather gloves as he pulls them on. "I could make more money somewhere else, but right now I want to do good things for people."

Outside the barracks, the sky is closer. It seems to have collapsed over Central Area like a gray dome covering the big quadrangle. The snow is piling up in drifts, blowing into the doorways. (No one knows it yet, but the long weekend holiday the cadets have been anticipating will be cancelled in the interests of safety. "Another weekend in my room," a senior writes the following week.) The snow muffles all sound except the crunch of boots. Then a bugle blows the repetitive "Mess Call," the notes sharp in the cold. All around, cadets, their shoulders hunched against the wind, move to their communal meal.

On the fourth floor of Bradley Barracks, Cadet Shawn Kilcoyne, commander of Company E-2, is talking his way into a decision Major Rob Olson says he'll regret.

"Tell me again why these people should go on pass?" Olson says, a tone of disbelief in his voice. Olson stands on one side of a conference table in his office, Kilcoyne stands on the other. Between them lies the company passbook, which records requests and approvals for weekend passes, and the "slug sheets," which record offenses and punishments under cadet regulations.

"Some of the cadets have major things going on this weekend," Kilcoyne says.

"And so you want to hold off on these," Olson says, fingering the slug sheets.

It is the Wednesday before a long weekend. Cadets in good standing can turn a weekend pass into three days away from West Point if they're proficient in all their subjects and not under restriction. To say that cadets are eager to leave West Point during the gloom period—the gray months between Christmas and spring leaves—would be like saying they lead a more structured life than students at Florida State. Kilcoyne is making a case for leniency.

"What things are going on?" Olson asks.

"One of these guys," Kilcoyne says, tapping the book, "his parents are having a twenty-first birthday party for him. The others have been invited. The platoon leader thought we should use discretion here."

"Tell Joey," Olson says, referring to the platoon leader, "and his discretion to come see me."

Kilcoyne shuffles his feet. He knows he's waffling, trying to pin the decision on the platoon leader. As the commander, he either supports the recommendation—and needs to say so to Olson—or he doesn't, in which case he should have had the guts to say so.

"I think we should use discretion, sir," he says.

Sergeant First Class Mercier, the company's Tac NCO, sits on a vinyl couch in the office. He looks like a man who is biting his tongue.

Olson uncaps a felt-tipped pen and writes on a pad: *Make the right, not the popular decision.* He underlines "right" three times. Kilcoyne looks at the pad, but he's steady on course.

"I thought that since there was, you know, parental involvement, and since you only turn twenty-one once, and it's a big birthday and everything, that we ought to let them go."

Kilcoyne is in a tough position. He's a new cadet commander (the semester is only nine days old). More importantly, he's dealing with cadets who are all but peers, and he's caught between them and the Holy Grail of cadet life: free time, the weekend pass, a few precious hours away from the little gray cell. Clearly he's decided it would be better to face Major Olson than to lay down the law in the barracks.

Olson walks behind his desk, but doesn't sit. Behind him, fat snowflakes fall steadily outside the window. He thrusts his hands deep in his pockets, stares down at the blotter, rocks back and forth on his heels. Kilcoyne stays next to the conference table, watching his Tac and squirming a bit.

Olson knows what's going on here: Kilcoyne is knuckling under to the pressure. And while a twenty-first birthday is a big deal, if he caves in he may never regain the respect of the rest of the company. Kilcoyne will be shoved this way and that every time some cadet has a "major thing" that he or she believes should take precedence over regulations, over strict and fair.

Kilcoyne has been asked to make a hard call. Every theory, every discussion of leadership he's ever encountered talks about the need to make tough decisions and live with the consequences, but he doesn't seem to know it.

"If leading soldiers has a degree of difficulty of five," Olson said over the summer, "Leading peers is a ten. But if I make the decisions for them, the cadets won't learn."

Olson isn't thinking about a canceled twenty-first birthday party or disappointed parents. He's trying to decide how to make the best use of a learning opportunity. How can he best teach Kilcoyne, this soon-to-be-lieutenant, that he must be fair in all situations? If Olson disapproves, chances are Kilcoyne will put it just like that. *Hey, I wanted to let you go, but the Tac slam-dunked me, man.* And he won't have learned the lesson.

If he lets Kilcoyne make the wrong decision—and any corporal in the Army knows that playing favorites with punishment is a bad decision—morale in the company will suffer. Cadets who are looking for reasons to be cynical about the system will have new ammunition. Olson does the math in his head, then looks up.

"OK," he says. "Your call."

Kilcoyne leaves the room, thinking he's gained a victory. He'll no doubt tell his classmates that he won a round from the Tac. As soon as the cadet is gone, Sergeant Mercier says, "Boy, he'll live to regret that decision. I mean, I can understand his reasoning, but it's going to make it impossible for him to get things done around here."

"That's the decision I expected him to make at this point," Olson says. "We have to make sure he learns from this. He's going to fall flat on his face the next time he has to make some unpopular decision and they all say, 'Hey, what about those passes?' "

Even Olson wouldn't have predicted how fast Kilcoyne's decision would come back to haunt him. A mere twenty-four hours after their meeting, another firstie is cited for a parking violation. When Kilcoyne tells his classmate the citation means restriction and loss of weekend privileges, the firstie reminds Kilcoyne of the birthday party favoritism. When Kilcoyne asks Olson what to do, the Tac gives him two options: Live with the lower standard, or get up in front of everyone and say, "I screwed this up."

Kilcoyne, looking for a less humiliating way out, insists that he is only trying to take care of soldiers. Olson tells him, "That's bullshit. What you did was sell out the system you're supposed to uphold. It's just like the cadet prayer says, a choice between the harder right and the easier wrong. You ignored that."

By letting Kilcoyne learn this particular lesson this particular way, Olson demonstrated that E-2 is in fact, not just in name, a cadet-led company.

"I'm running a little experiment this semester," Olson says. "It could be great or I could fall on *my* face."

Kilcoyne is part of the experiment. Olson picked his "front-runners," his top performing cadets, for chain of command positions the first semester. This semester, he chose leaders from the middle of the pack. He wanted to see if he had created a developmental environment for all of the cadets in his company.

"In the first nine days [of the new semester] they've done more things right than my first semester chain of command."

Olson opens his desk drawer, pulls out a can of smokeless tobacco and puts a pinch beneath his lower lip. He finds a Coke can and carries it to the table with him. He punctuates his sentences by spitting into the can.

Olson concedes that a lot of Tacs—maybe most—wouldn't have handled Kilcoyne that way. There is a danger in letting his cadet commander make the wrong decision: The other cadets in the company will see the injustice—and that will be the lesson they take away.

"I've got to make sure the CO's lesson is very public—he's got to fall down and skin his nose without losing all effectiveness—and I've got to sponsor that."

Olson believes that leaders must learn to deal with failure; and if they don't fail on their own, he isn't beyond throwing a little disruption in the way to force them to react, force them to handle the confusion. He did that with Kevin Bradley during Beast the previous summer.

Bradley, he says, had the job "wired" in a couple of days. The company was running well, but Olson saw the learning curve flattening out because things were going smoothly. One day the company finished training early and was all set to eat chow, then meet a truck convoy for the ride back to the barracks. It was a smooth operation right up until Olson called the operations center and asked an officer to have the trucks sent out early, just as the company was about to eat.

"I said to him, 'Don't tell anybody I called you,' Olson says, smiling and spitting into the can. "Sure enough, the company is about to start eating and the trucks show up and now Kevin has to deal with the whole mess."

Months after that summer night, Bradley—who didn't know the foul-up was his Tac's handiwork—said it was a stressful evening. The cadets had set up the chow line in the only space available for the vehicles to turn around. The huge trucks backed up on a narrow mountain road, belching diesel fumes. A hundred and fifty hungry new cadets and thirty cadre members stood in platoon formations on a path leading from the training site; they could smell the food from where they stood in their neat ranks. Four or five officers, a half dozen NCOs, a dozen truck drivers and their NCO were all crammed into this small clearing between the base of the hill and the road, and all eyes were on Kevin Bradley.

Some of the cadet chain of command wanted to pack up the meal and send it to the rear and unpack it and set it up all over again. Others wanted to rush the new cadets through the chow line. The sergeant in charge of the trucks and drivers was clearly agitated that he and his troops were being jerked around. He didn't want to wait, and Bradley didn't want to piss off the transportation chief.

And the whole mess had been created by Rob Olson as a training vehicle for his company commander.

"Kevin ran around a little bit, but then he made his decision. He had the new cadets eat, though he gave them less time. He briefed the NCO with the trucks and kept everyone informed. He wasn't quite sure about it all, but I asked him if he was accomplishing the mission. He said yes and even figured out that they would still get back to the barracks ahead of schedule, so the squad leaders would have a little extra time."

"The only time you grow as a leader is when you get outside your comfort zone," Olson says. "When you're not the master of your universe. I'm not all that worried about my boss seeing me make a mistake, because I think he'll find me doing nine things right for every mistake. I tell the cadets that there are going to be days when you have shitty training. If your boss comes to see you on that day, you'd better be able to say, 'Sir, I grew some privates and NCOs who now understand how to make the Army training system work.' "

There is no handbook for Tacs; they are charged with developing their subordinates into leaders. They take the lead in the Academy's trickiest mission, and there is not even a consensus on how it's to be done. If Olson has 150 cadets in his company, he will keep 150 different development plans going at once. Not every cadet needs the same thing, and not every cadet has the same experience. Some will screw up, some will have an easier time and not learn as much as those who do fall down. Not all of them will get to be platoon sergeants and company commanders. Yet they all must be ready on graduation day. Much of the responsibility for their preparedness falls on the tactical officer.

A second class cadet knocks on the door—two sharp raps—and asks for Olson's signature on a form that will allow her to overload her courses this semester. He asks about her schoolwork, then brings up a plebe in the young woman's squad. The plebe has been having trouble handling things, and the upperclass cadets do not know quite what to do about her.

"You can't hold her hand," Olson says. "If she wants to go into the latrine and bawl her eyes out and wallow in self-pity, let her. But I get the sense the upper class backed off."

The squad leader nods. It's one thing to lean on plebes who screw up but are still trying to do the job. It's quite another—it's even perceived as dangerous—to lean on a plebe who is having a meltdown. Olson insists that the plebe must be treated like her classmates.

"Ask yourself: Is she going to be ready to stand on a street corner in Bosnia in a few years?"

He spits again, then puts the can down on the table as he signs the young woman's form. A copy of *Parameters*, the professional journal of the U.S. Navy, is open on the table. Olson is reading an article titled "Military Values and Ethics." Beside that is a book: *Breaking the Phalanx: A New Design for Landpower in the 21st Century*. On the cabinet that holds the sink there are three small toy cars, either left over from a visit by his son, or waiting for the next visit. The cars, the constant stream of cadets, the professional reading program—Olson's life is a juggling act just as the lives of his cadets are juggling acts.

Olson has been selected for the next level of the Army's professional schooling, a year at the Command and General Staff College at Fort Leavenworth, Kansas. C&GSC, as it is known, is another gate, another bottleneck, another in the narrowing steps up the pyramid of promotion. Olson's selection, along with his early promotion to major, are indicators of his success, but he isn't sure he wants to go because it might mean a year away from his family. His wife, Major Holly Olson, an obstetrician at West Point's hospital, may have to stay.

"They need an OB/GYN at Leavenworth," Olson explains. "Even tired old majors and lieutenant colonels out there still fornicate and make babies. But we're not certain Holly could come along."

The Olsons have requested what the Army calls "joint domicile," meaning they would be assigned to the same installation and could continue to live as a family. Even though Olson says that the Army does what it can to keep doctors happy (they're hard to retain), there is no guarantee the Olsons can stay together.

There are many reasons to stabilize assignments, to keep soldiers at one installation for a planned three- or four-year tour of duty: it makes personnel planning easier, more predictable for the service; it helps reduce the unavoidable turmoil in units when leaders and specialists, such as Dr. Olson, leave. Most importantly, it saves money.

"The near-term battle is to get Holly to Leavenworth. If we can't do that, she and the kids will head out to Fort Riley [also in Kansas, several hours drive from Fort Leavenworth] and I'll get an assignment there after school."

Hands behind his head, Olson thinks about what the summer might hold for him and shows the first chink in the nearly faultless armor of the happy, self-actualized Army officer. "This could be tough on the kids," he says.

Olson's apparent ingratitude at being selected for schooling wasn't received well in Army headquarters. He called and said he didn't want to be separated from his family for a "bunch of stuff" he could learn from books.

"General Miller, one of my old bosses, got wind of my little temper tantrum and called me up. He really chewed my ass out. Told me to act like an adult, to ask for help before I start making stupid threats. I didn't even get to talk, I just listened. And he was right, of course."

Despite the threats, Olson is happy in the Army and unsure of what he would do as a civilian. Although classmates now in business have approached him about getting out of the service, he has a limited understanding of what civilian managers do.

"I love the Army," he says. "I'll truly regret leaving when the time comes, but the Army doesn't define me. It's a thing I do. My family, my relationship with my wife, they define me."

Olson's comments sound a lot like the aphorisms with which the cadets like to decorate the barracks. But he lives this way; it is the source of his calm demeanor and confidence: He has found a job he loves, he has taken the time to figure out what's important to him.

In 1987, when he graduated, it didn't look to Rob Olson that he was on track for any of this. Academically, he did not succeed at the Academy, attending summer school three times to make up course failures.

"By all standards West Point had I was going to be a shitty lieutenant. I was a poor student, although I did well physically and in my military grades. But I wasn't that confident coming out of here. Then I got to the Army and realized nobody expects me to memorize Newton's laws. What they wanted was someone who was sincere, who had common sense, who had enough charisma to make a unit come together."

He attributes a lot of his success to good mentors he had early on, some of whom, he admits, chewed him out constantly. He also learned that while it was important to succeed in the bosses' eyes, it was just as important to be judged a success by his soldiers and NCOs.

It is nearly lunchtime. Olson calls Holly and asks if she has time to meet for lunch. "Maybe she won't deliver until after lunch," he says into the phone. "Tell her to cross her legs."

"Holly always loves when I give her medical advice," he says after hanging up.

Outside, it is still snowing. The sky seems to be falling, and the whole world is done in black and white. Olson walks up the hill behind the barracks to an access road where cars and trucks are parked in a tight line. He pulls a set of keys from his pocket and unlocks the door to a huge, bright red pickup truck, which was a combination Father's Day, birthday, and Christmas present for a couple of years.

The truck seems impractical: There are four people in the Olson family—too many for the bench seat. Olson doesn't have to haul cargo in his job as a mentor to cadets. But just as country music and gun racks are part of the landscape in GI towns like Columbus, Georgia (outside Fort Benning) and Kileen, Texas (outside Fort Hood); trucks are part of the culture for Army officers and NCOs.

Keller Army Hospital sits at the north end of post, in the shadow of Storm King Mountain. Unlike the other buildings on West Point, the hospital wasn't designed to look like a medieval fortress. It is sleek and modern, with long banks of smoked glass windows and a pleasing, symmetrical shape. It is, however, gray.

Olson greets the soldiers and civilians who work in the OB/GYN clinic. Like the cadets he speaks to everywhere he goes, they all seem to know him. He sticks his head into a small office where a woman in battle dress uniform sits filling out a logbook, her back to the door. "I'm here for my checkup," he announces.

"Great," she answers without turning around. "Feet in the stirrups and I'll be with you in a minute."

Like her husband, Holly Olson has the easy physical grace of an athlete, a relaxed demeanor and a firm handshake. Five nine, with wavy brown hair cut short, round glasses that make her look like what she probably always was in school: the smart girl in the class. She is unguarded and engaging, a natural conversationalist.

They head to the hospital cafeteria for take-out. The food here, as in every subsidized restaurant on-post, is cheap and plain. After collecting a raft of sandwiches, chips, and sodas, the Olsons head back to Holly's office. The room is small, about twelve by twelve, with a window high up on one wall, two large desks, a chair for the patient and a tiny sink in the corner with bottles of pink hospital soap. There is a spray of flowers in a vase on Holly's desk, long branches with tiny buds that reach almost to the writing surface. Rob sits in the patient's chair beside the desk, holding his sandwich box in his hand. Holly sits at her desk, turned sideways. Their feet—they are both wearing black combat boots—are toe-to-toe.

Olson is his wife's biggest fan.

"She'll jump out of bed in the middle of the night when her beeper goes off, come down here and save the lives of baby and mom in some medical drama. I spend my day worrying whether or not Jocko's locker is ready for inspection, yet when I come home she listens to me as if I have the most interesting job in the world."

"It's hard," she tells him. "You've got a lot to keep track of."

Things haven't been easy for this couple.

After graduation in 1987, Rob Olson was sent to Fort Sill, Oklahoma for his officer's basic course in the artillery school. Half a year later he joined his first unit, the famed 101st Airborne Division, at Fort Campbell, Kentucky. Holly Louise Hagan, also USMA '87, resigned her brand-new commission (with the Army's blessing) to attend Medical School at the University of South Carolina. Olson drove back and forth to South Carolina to visit Holly, who was a busy first-year medical student. They continued to date long distance, with Olson making the several-hundred-mile drive as often as his schedule of field problems would permit.

They were married in 1989. Later, Holly applied to medical school at Vanderbilt University so she and their new daughter would be closer to Rob.

"So there Holly was," Olson says. "Going back to medical school after a year off [for maternity leave], worried about how much she'd forgotten. Oh, and she's in a new school in a new city and doesn't know any of her fellow students. And she has a new baby. And the nanny we'd arranged for when she got to Vanderbilt quit right before Holly got there."

It was just at that time, when it seemed as if things couldn't get any more stressful, that the 101st Airborne Division deployed to the Persian Gulf for Operation Desert Shield.

"First letter I get from Holly, I'm sitting out there in the desert and I read about the nanny quitting and the new school and the new baby and it sounds like she's about at the end of her rope. And I'm in the middle of nowhere. I told my battery commander, 'Sir, I'd like to call home.' And he said something like, 'Hey, Rob. Use the first pay phone you see.'

"It was forty days before I could call her," he said. "By that time she had everything under control, of course. But I had forty days to wonder how she was doing, and there wasn't anything I could do to help."

Olson laughs as he tells. He never doubted that Holly would pull everything together; it was about him wanting to help her through a tough period.

One wall is decorated with a large photograph of Tripler Army

Medical Center in Honolulu, where Holly served for several years. The big pink hospital sits on a hill overlooking Pearl Harbor. The matting of the photograph is signed with good luck messages from the hospital staff. Where Rob's office is decorated with guidons and flags and plaques, Holly's has this one photo.

She did not come to West Point to become a doctor, but during her first two years as a cadet she discovered two things: she wasn't interested in the branches of the Army most West Point graduates join, and she found that she was still a top-performing student. Chemistry was her favorite subject, but there are no jobs for chemical engineers who are second lieutenants. She says she became a doctor by default. "I'm still not sure my dad believes I'm a real doctor."

Her parents were supportive, but her home was no hotbed of feminism. Nor did her family have any ties to the professional military. Instead, she looked around her home town and saw all kinds of people with bachelor's degrees and no job, people still living with their parents. She didn't want that to happen to her.

As a cadet, Holly dated a few classmates, but it didn't work out, for the same reasons the women cadets in the Class of 2000 mentioned: the small town atmosphere, where everybody knows everybody's business. And for Holly, who graduated near the top of her class, she had to deal with the fact that many young men are intimidated by smart women.

"I didn't wear it on my sleeve or anything," she says. "It's not like I walked in the room and said, 'Hey, did I tell you today that I'm smarter than you?' "

Holly finished seventeenth in their class, on her collar she wore the gold stars that mark the top cadets. Her husband was ten from the bottom and a veteran of summer school, called STAP (Summer Term Academic Program). He characterized their dating as "STAP-boy dates Star-girl."

"Rob was persistent," Holly says. "He kept asking me out and I'd say 'no' and he'd ask again and I'd say 'no.' Then one day I said 'yes' and he was really surprised."

"I already had plans for the weekend," Olson adds. "Because I was just expecting to get shot down again. I was going down to Fordham to watch some friends of mine play rugby. Then she said yes, and I had already told my buddies I'd be there, so I asked Holly if she'd like to come along. As soon as I got my car from the parking lot it started to pour."

"Hard," Holly adds. They're a team now, telling a story they both know well and have laughed about before.

"And of course neither of us has a raincoat," she says. "I was wearing this big wool sweater."

"So I stop at this convenience store," Olson says. "I have about seven bucks on me. So of course I buy a six-pack of Bud and a box of plastic trash bags to wear as rain gear."

"We got drenched," Holly says. "And it was freezing. We smelled like dead sheep."

"So after the game I decide I'm not bringing my first date to the rugby party—I was smart enough to spare her that. Instead we're going to drive to New Jersey."

"We were going to Joey Simonelli's parents' house," Holly adds.

"Right. To dry off." Rob looks at me. "Joey is a classmate and one of my best friends. See, I was making this date up as I went along."

Holly lifts her hand to her face as she laughs.

"And we're getting close to the George Washington Bridge, and I realize I spent all my money on the beer and the trash bags. I've got no money for the toll," Rob says.

"Here I dragged this girl out into the pouring rain to watch a rugby game, and now I didn't even plan well enough to have the bridge fare. So I know I have to ask her for money, but I'm such a coward that I wait until we're on the ramp—there's no turning back—and I blurt all this out. And she just looks at me . . ."

Holly finishes the story. "I just looked over and said, 'There's no toll in this direction.' "

They share a laugh, and Olson says, "I exposed my ignorance and I didn't even have to."

"But there I was with this guy I'd been avoiding, and everything that could go wrong was going wrong, and I was having a great time,"

Holly says. "We laughed the whole time and we had interesting conversations. It was a great date after all."

Outside, the after-lunch appointments are gathering: young women, some of them in uniform, with swollen bellies. They clean up the plastic trays, and Holly washes her hands in the small sink. After exchanging a few notes about their schedules, they plan on being home for family dinner.

A couple of weeks later the Olsons are guests at a banquet, a formal dinner in the Mess Hall, hosted by the sophomores for Yearling Winter Weekend.

Rob Olson wears a wasp-waisted coat, dark blue, and light blue pants that make up the officer's dress mess uniform. The lapels of his jacket are faced in scarlet, the color of the field artillery; the shoulders are topped with gold braid. It is the kind of dashing uniform that looks good on young men; it also sports enough baubles and brass and gold to make it an easy target for caricature. Holly Olson wears a long black dress, appropriate for the drafty Mess Hall. All around, second-year cadets, their dates, and their families glitter and shine. Many of the yearling women are escorted by firsties; most of the yearling men have civilian dates.

This is just one of a string of formal dinners the Olsons will attend as part of their official duties: there was Plebe Parent Weekend in the fall, Yearling Winter Weekend in January, followed by One Hundredth Night (one hundred days until graduation) for the first class, and Five Hundredth Night (for the second class).

"I could just keep the monkey suit on a hangar and jump in it every weekend," Olson says.

"And the menu, believe it or not, has been the same for each formal," Holly adds.

But they are smiling. The other Tacs, most of them are men, and their wives, come up to the Olsons. The young couples are taking advantage of having baby-sitters at home, and are heading to the South Gate Tavern, just off post, for a few drinks. Everyone seems in fine spirits—the women are beautiful, the men are handsome.

"You going out dressed like that?" Sergeant First Class Tim Bingham asks Olson, pointing to the glittering uniform.

"I can shoot pool in this as easily as in jeans," Olson says. Then, leaning forward so that only Bingham can hear, "And I can say, 'Hey, get off my stool, you rednecked motherfucker,' just as easily in this." He straightens. "Matter of fact, it makes me look taller," he says pulling at the bottom of the jacket and smiling his mischievous smile.

Everyone is smiling, in spite of the weather, which is frigid, in spite of the fact that the semester has just begun and these cadets are not yet halfway through their time at West Point. On this night they shine. Cadets stand outside the little circle of officers, waiting to introduce parents to their Tacs and Tac NCOs. Mothers and fathers beam. Young women in slinky gowns try not to shiver as the big doors of the Mess Hall swing open and a January wind spills inside. Many of the women—the ones with gallant and resourceful dates—wear cadet parkas over their evening dresses. The tables are set with cloth napkins; the steel flatware has been polished; the peanut butter has been removed from the table.

Olson jokes with his fellow Tacs, and although they are a little old to be fraternity brothers, there is good-natured shadowboxing. The Olsons have made it to this place through hard work. They have been successful in the Army, and they are destined for more success. They are surrounded by friends, by work they love. They live in a tight-knit community and mostly overlook the downsides of the small-town atmosphere. And no matter how tedious it will seem on Monday morning, when Olson is once again consumed by thinking about whether Jocko is going to be ready for inspection, or whether some procrastinating firstie has bought his officer's uniforms, or whether some plebe is crying in the latrine at night, tonight it is like a fairy tale. Or as much like a fairy tale as real life can get when the baby-sitter is waiting. They are in their anointed season, young and strong and healthy, with everything to look forward to.

TRUST BETWEEN LEADER AND LED

The notion that failing—and the learning opportunity that follows—must be part of leader development is not just a heretical idea held by a few social scientists toiling away in the windowless offices of Thayer Hall.

"You have to be able to fail and learn from it," Brigadier General John Abizaid, Commandant of Cadets, says. "I talk about my screwups; God knows I have plenty of material. People have to know that they can learn from their mistakes. We do ourselves a disservice with the idea that we want people who don't fail."

He is in his office above the Cadet Mess. In the daylight, the Plain and the river beyond are visible through the arched windows. On this winter evening, the barracks lights throw yellow rectangles on the snow; General Washington, astride his horse and below the windows, looks out over the quiet cold.

Abizaid hangs his camouflage field jacket on the back of his chair.

He is forty-seven, dark-haired and handsome, with a friendly smile and a mischievous sense of humor that doesn't seem to fit his role as head of the military side of cadet life. He graduated forty-second (of 944) in the Class of 1973, studied at the University of Amman in Jordan, and at Stanford; he also has a master's degree from Harvard.

But Abizaid has not hung his Harvard degree, or any other diploma, on the walls of his office. Instead, he has two paintings of 82nd Airborne Division soldiers in World War II combat. One shows a general officer leading from the front: Matthew Ridgway, USMA 1917, in a battle for the LaFiere causeway in Normandy, June 9, 1944. The two-star general has moved to the very front of his stalled attack and is personally exhorting his soldiers to press on.

Abizaid walks through the warren of hallways above the Mess Hall to an amphitheater-style classroom filled with forty or fifty cadets of the Infantry Tactics Club. These cadets, and there are men as well as women, spend some of their precious free time studying and practicing the Army's most basic craft: how to close with and destroy the enemy.

The cadets jump to attention when Abizaid enters. He greets them in his breezy style, then takes his place in front of the room beside an overhead projector. Using colored markers, he sketches a map. There is a small hill at the top of the frame, overlooking an airstrip that runs left to right across the center. South of the airfield, some blue lines denote a body of water. To the right of the airfield, on the east side of the map, Abizaid draws a half dozen buildings, little black squares and rectangles. He labels them "CAMPUS."

"There are American civilians here," he says, pointing to the buildings. "Our job was to get them out safely. The plan looked like this."

He puts down a second sheet, an overlay on which he draws military graphics that show how a company of infantry was to move off the airstrip, secure the buildings and the civilians, then prepare for a counterattack from the east, where the enemy had light armored vehicles.

"That's what everybody thinks is going to happen," he says. Then he pulls off the overlay with the original plan and tosses it to the floor with a flourish.

"But we need to get rid of that because, of course, the plan never works."

He begins to speak more quickly, using a red pen to draw enemy forces. There are powerful anti-aircraft guns on the hills above the airfield that can be turned on the GIs with devastating effectiveness; these same guns will also keep the planeloads of American reinforcements away. There are several dozen armed policemen among the American civilians on the campus; the enemy light-armored vehicles are not a forty-five minute drive away—as reported—but are much closer. And the construction workers who'd been laboring on the airstrip are armed and organized into a company of infantry.

This is Grenada, 1983, and the American civilians are medical students. The cadets in the room were only a few years old when this combat action took place. John Abizaid was a captain, ten years out of West Point and commanding a company of about a hundred Army Rangers.

He speaks to the cadets as if they are in command of the men on the ground.

"You underestimated the enemy and what they could do to you," he says, looking into the audience. "There's no real plan for fire support. The BTR-60s [Soviet-made light armored vehicles] are much closer than you thought. You're suddenly taking fire from this hill. What do you do?"

"Attack the hill," several cadets respond. This may be natural aggressiveness, or they may be showing off for the Commandant. Perhaps a few of them learned this in a military science class.

"Of course," Abizaid says. He shows how the company commander changed the plan. Two of the company's three platoons now move up the hill to silence the powerful enemy guns, leaving one platoon to take on the police, secure the medical students, and protect the company's flank from a mounted assault.

"Now, you're this platoon leader," he says, using a pointer to show the soldiers moving off by themselves to the east, toward the expected counterattack. "What's your first concern?"

There is a chorus this time. "Security," the cadets answer. This is one of the principles of war, drilled into them in military science and military history and plebe knowledge. On the ground it translates to, "expect the unexpected."

The platoon leader given this mission, Abizaid points out, has been with this unit only three months, has been in the Army less than a year and a half. He pauses to let that point sink in.

This is all much closer to you than you think.

"The whole United States of America is going to be watching him and what he does to protect these civilians."

The cadets going into World War II probably never heard this admonition. But the young men and women in this room have grown up with twenty-four-hour news; they know that CNN might swoop down on them at any time, putting them—and their actions—on the world stage for everyone to judge.

The Commandant walks them through the battle, and the cadets are riveted. Abizaid, a natural storyteller, gives them the sights and sounds, the heat, and the worries of command. He has conveyed the shock of what happened that day: the GIs on the ground had been in their own barracks just hours before. Many of them know they're on an island called Grenada, but because the maps they've been given show only the island, they don't know where Grenada is.

The lieutenant and his platoon of forty-some men secure the buildings of the medical school and set up positions facing east, waiting for the mounted attack. On the hill, GIs are dying in the assault on the big guns.

"Now what do you do?"

They are not as quick to answer now, but one cadets ventures, "Send out OPs [observation posts]."

Abizaid taps the map with a pointer. A large hill to the east keeps the Americans from seeing what's on the flank, so the lieutenant sends a jeep, just landed off an aircraft, out along the road heading in that direction. Then he hears gunfire, and the jeep doesn't return.

"The company commander tells the lieutenant to get over to this hill and find out what's out there."

The lieutenant spreads his platoon too thin, then compounds the problem by walking away from his other leaders; he is now out of contact with the unit he is supposed to command. When a GI destroys an approaching armored vehicle with an anti-tank weapon, the lieutenant—who is now farther east than any man in his platoon—goes out on the road to inspect the damage.

"Good idea?" Abizaid asks.

"Bad idea," a cadet responds.

"Right. But let's not forget that it's easy for us to sit in a classroom and criticize this guy," he reminds them.

"He got shot five times. One of his guys had to crawl out there and drag him back."

The entire action took less than four hours. In that time, Abizaid points out, the lives of soldiers and civilians rode on decisions made by that lieutenant.

Abizaid points to a first class cadet in the front row and directs him to a white board in the front of the room.

"We could put these lessons in terms of the Principles of War, but for our discussion, let's just do them this way."

"Never underestimate the enemy," he says. The firstie writes.

He shows them another map, a sketch of the battle at the Little Big Horn. George Custer, who took several companies of cavalry to their deaths along that river in Montana, lies buried in the West Point cemetery less than a mile from where these cadets sit.

With a little prodding, Abizaid extracts a few more lessons.

"Never send soldiers somewhere they can't get help. Never plan for a fair fight."

The cadets nod; some of them take notes.

"Now, here we are in Bosnia," Abizaid continues, pulling the map of Grenada off the projector.

Many of the cadets in this room know recent graduates on duty with the peacekeeping force; some of the seniors expect to join that force within a year. And if that is not enough of a reality check: the day's newspapers are filled with the story of American troops being sent to Kosovo on yet another peacekeeping mission.

"We have this Muslim village on the boundary established by the Dayton Peace Accords."

Abizaid puts up another slide that he has prepared ahead of time.

"There are supposed to be no weapons in this village or in the boundary area. The Muslims who lived here before want to move back into the village. A report arrives at American headquarters that the former inhabitants want to go home and rebuild. There is also a Serbian complaint that the Muslims are moving weapons into the town, that they plan to use it as a staging area for 'terrorist activities.' "

A young lieutenant, a recent West Point graduate, gets orders to check out the Serbian complaint and ensure that no weapons are coming into the village. This is exactly the kind of scenario Abizaid is preparing these cadets for: quick thinking, handling diverse cultures, with a mission that is anything but simple. Compared to this kind of mission, an order to storm the village and kill everyone inside would be easier to figure out, if more dangerous.

"You're the lieutenant," Abizaid says. "You've got forty-eight hours to prepare for this mission. You know that the roads are clear, because UN vehicles have been using them, but other than that there are lots of mines and unexploded munitions all over the area. You get a few translators. What do you do?"

Abizaid sits on the front edge of the instructor's desk, dangling his feet and holding the pointer in two hands.

"Before you get out your Ranger handbook or start asking for checklists, let's just spend a few minutes thinking about what kind of things you'd like to know."

The cadets, eager to participate, begin to call out their concerns.

"Where are the nearest friendly units?" one asks.

"That's right," Abizaid says, referring to the note on the board.

"Don't expect a fair fight. If they are terrorists and they start shooting at you, you've got to be ready."

The cadets warm to the task.

Where are the nearest reinforcements, they want to know, and how long would it take them to get there? How do I get to them on the radio? Who's in command? Can I talk to the commander? Can we do a reconnaissance? Can we get a helicopter to fly us over the area?

"Good, good," Abizaid encourages them as they think through the problem.

"How many of you studied what happened in Mogadishu?" he asks, referring to the fierce battle, in October 1993, between elite American forces and Somali militias.

Twenty hands go up.

"Part of our problem was that we went in there thinking these were a bunch of dumb tribesmen. But they turned out to be a bunch of smart tribesmen. They sat around every day watching the Americans and thinking about how they could kill our men, how they could embarrass us in front of the whole world. Don't you think the Serbs and Muslims, who've been killing each other for seven hundred years . . . don't you think they're thinking about how they can kill you?"

The problem isn't that American forces don't have the weapons or the training; the problem, Abizaid says, is a mental letdown. The problem is complacency.

"There are captains and lieutenant colonels and old generals who will tell you, 'Lieutenant, it's a routine mission, there's no need to get all worked up about it.' I'm telling you right now . . . that's bullshit."

There is a silent moment. The cadets sit in their rows, upright and attentive. In a year or two, some of them will be in this position, in Bosnia or Kosovo or some other country whose name they do not know today. The Commandant sits on the desk, pointer in his hand, feet swinging gently back and forth. Because he's been there, he can picture them as they will be: K-pot crammed down, rifle slung over one shoulder, dirty notebook and GI pen in grubby fingers, looking for the answers that will accomplish the mission and keep their soldiers safe. Abizaid is passionate about preparing these young men and women as well as he possibly can.

"So what happens?" Abizaid says, standing and picking up a marker. He draws a couple of rectangles on the road leading to the village.

"The platoon moves out in column . . ."

There are groans from the audience; the cadets know that vehicles in a tight line cannot protect themselves or each other.

"The lieutenant sets up a road block right in the middle of town, down here in the low ground."

The roadblock is ineffective; the soldiers manning it cannot see more than a block in either direction. Their radios can't reach over the surrounding hills, which means they can't call for help if they need it. The other vehicles are set up in places where they couldn't support one another if they came under fire. It's Custer, splitting his column, the pieces too far apart to help one another. It's his platoon leader in Grenada, wandering out in the road to inspect his kill, far beyond where he could control things.

"Turns out the Muslims were smuggling weapons in, and the Serbs were preparing to attack. But this platoon leader didn't know it because he wasn't set up properly. Fortunately his company commander came down and saw what was wrong and got everything straightened out and peace prevailed."

Abizaid turns off the overhead projector and picks up a photocopied article from a professional journal.

"Leadership is the most dynamic aspect of combat power," he reads. He looks up. "No technology can give you the advantage that good old-fashioned leadership can give you. You can have spy planes overhead and all kinds of information downloaded to the G-2 [the intelligence section]. If the lieutenant and sergeant on the ground don't do their duty, we will fail."

"When you hear stories like this, you should have a couple of reactions: First, you should think, 'But for the grace of God, that could have been me.' This lieutenant [in Bosnia] was lucky. His company commander was looking out for him."

"Second, you should read about these disasters because you don't want it to happen to you. You don't want them writing a book about *your* big screwup."

"Your job," he tells them, "is to wargame, to think through what might happen. We want you to be tactical leaders, to avoid the mistakes of the Little Bighorn, of Grenada. If I had given that lieutenant [in Grenada] clearer instructions, maybe he wouldn't have gotten shot. I have no excuse. I take responsibility."

When the meeting is over, the cadets head back to their rooms. Abizaid rides the elevator to the first floor, then steps through the big front doors of Washington Hall and into bitter February cold. Big lights on the barracks paint the scene white and gray, making everything look colder.

"I'm passionate in the belief that we train lieutenants to fight our wars," Abizaid says.

There are dozens of cadets, most of them in sweat suits, hustling by. They salute the Commandant.

"Are you staying out of trouble, Joe?" the Commandant calls to one cadet he recognizes.

"Sir, I'm out looking for it."

"What did you wind up choosing, Joseph?" Abizaid asks. The cadet is a firstie; the Comm wants to know about his recent choice of a first duty assignment.

"The eighty-deuce, sir," comes out of the night air.

Abizaid continues to walk, a small smile on his lips. He served two tours with the 82nd Airborne Division, including his first assignment out of West Point. "I'll alert them that you're coming," he tells the cadet.

The six floors of MacArthur Barracks loom above him as he walks, the windows glowing brightly; it is the beginning of study barracks, and everything is quiet. No stereos playing, no music coming from the building. There is no rule forbidding music during evening study period, but each company is responsible for enforcing good study conditions.

"Preparing for war doesn't just mean we do military stuff all the time," Abizaid continues. "We need flexibility, people who can think. One criticism I have of West Point is that we turned out automatons. We were famous for our rigidity. That's not good, not in Custer's time, certainly not in the Bosnian situation I just described."

He passes the statue of MacArthur and, in a few steps, is across the street from the Superintendent's quarters; the big windows of the old house glow warmly.

"None of these changes would work without the Superintendent," he says. "He inspires people. They *want* to do well for him."

This is, of course, the attitude the cadets have about the Comm, the same attitude the new cadets of Alpha Company had toward Greg Stitt and Grady Jett. It is part of the formula. The other part is the team, and John Abizaid will address that in the morning.

At 6:25 the following morning, the sky above the eastern mountains is pearly with the cold. Ice covers the sidewalk leading to Arvin Gym. Out front, a sign notes the hours: it opens at 0515. There are already scores of people about, though the Corps of Cadets is at breakfast.

The chlorine smell points to Crandall Pool, the home of Army swimming. There is an Olympic-sized pool in one end of the huge room and a diving pool at the other. The floating bulkhead that divides the two has been lowered this morning. In an overheated, glassed-in room that overlooks the pool lay dozens of gym bags, towels, athletic shoes, and Army PT sweat suits. The owners of all this clothing—the staff of the United States Corps of Cadets, the nearly one hundred people who work for the Commandant—are in the pool.

The PA system blasts rock and roll. Above that thunders the voice of this morning's instructor, Captain Carol Anderson. She wears the black shorts and gray shirt of the Department of Physical Education; a headset microphone is clamped over her short hair.

"Get off the side of my pool!" she calls, her voice absurdly amplified. "Tread water, tread water, tread water!"

The heads visible above the churning water push away from the side. Most of them are smiling; a few of them look frightened. There are a few flotation vests, visible just below the surface of the water, on the weak swimmers.

Brian Turner is with the other tactical officers. Without his glasses, Turner squints as he looks around. The whole scene shakes with the music as Joan Jett belts out her love for rock and roll.

"Hands up, hands up!" Anderson calls out. "Run in place!"

Hands come out of the water, some higher than others. A lot of these people are outstanding athletes, and they can't help showing off. The Tacs splash one another like kids at the neighborhood pool.

This is unit physical training. The Commandant calls all of his people together a few times a year, not because they need more exercise than they're getting, but to reinforce the sense that they belong to a team. There is a danger, as they work in their widely dispersed offices, that they might forget that crucial point. Here, in the pool, all they have to do is look around.

As Garth Brooks twangs, "Ain't comin home 'til the sun comes up," Sergeant First Class Tim Bingham, a broad smile on his face, laughs with his mouth wide open.

In this culture, if it's supposed to be fun, if it's supposed to be good for the organization, it's probably built around physical activity.

All the team-building has immediate results: The academy runs better when everyone has the sense of working toward a single purpose. But running a good school isn't the ultimate end, and there are reminders of that everywhere.

On the wall of the room overlooking the pool is a large frame with several photographs and a long piece of text. One shot shows Cadet Paul Bucha, '65, in his graduation photo. Another shows Bucha, captain of the Army swimming team, kneeling by the pool, all smile and lean muscle. Another picture shows Captain Bucha in dress uniform, the Medal of Honor hanging from a sky-blue ribbon around his neck.

Paul Bucha won the Medal of Honor for combat actions over a three-day battle in 1968, near Phuoc Vinh, Vietnam. The official citation, written in the stilted, artificial style of awards, compresses the events of three remarkable days into a couple of paragraphs.

Captain Bucha distinguished himself while serving as commanding officer, Company D, on a reconnaissance-in-force mission against enemy forces near Phuoc Vinh, Republic of Vietnam. The company was inserted by helicopter into the suspected enemy stronghold to locate and destroy the enemy. During this period Captain Bucha aggressively and courageously led his men in the destruction of enemy fortifications and base areas and eliminated scattered resistance impeding the advance of the company. On 18 March, while advancing to contact, the lead elements of the company became engaged by the heavy automatic weapon, heavy machinegun, rocket-propelled grenade, Claymore mine and small-arms fire of an estimated battalion-size force. Captain Bucha, with complete disregard for his safety, moved to the threatened area to direct the defense and ordered reinforcements to the aid of the lead element. Seeing that his men were pinned down by heavy machinegun fire from a concealed bunker located some 40 meters to the front of the positions, Captain Bucha crawled through the hail of fire to single-handedly destroy the bunker with grenades. During this heroic action Captain Bucha received a painful shrapnel wound. Returning to the perimeter, he observed that his unit could not hold its positions and repel the human wave assaults launched by the determined enemy. Captain Bucha ordered the withdrawal of the unit elements and covered the withdrawal to positions of a company perimeter from which he could direct fire upon the charging enemy. When the friendly element retrieving casualties was ambushed and cut off from the perimeter, Captain Bucha ordered them to feign death and he directed artillery fire around them. During the night Captain Bucha moved throughout the position, distributing ammunition, providing encouragement and insuring the integrity of the defense. He directed artillery, helicopter gunship and Air Force gunship fire on the enemy strong points and attacking forces, marking the positions with smoke grenades. Using flashlights [while] in complete view of enemy snipers, he directed the medical evacuation of three air-ambulance loads of seriously wounded personnel and the helicopter supply of his company. At daybreak Captain Bucha led a rescue party to recover the dead and wounded members of the ambushed element. During the period of intensive combat, Captain Bucha, by his extraordinary heroism, inspirational example, outstanding leadership and professional competence, led his company in the decimation of a superior enemy force, which left 156 dead on the battlefield. His bravery and gallantry at the risk of his life are in the highest traditions of the military service. Captain Bucha has reflected great credit on himself, his unit, and the United States Army.

This citation is one of the few complete versions displayed prominently at West Point, but the cadet area is filled with bronze tablets commemorating other Medal of Honor winners. They are fixed to the barracks walls, to the library, to Eisenhower Hall. They are like the statues of famous men in that they honor heroes, but they are different in a profound way.

The cadets who walk past Eisenhower's statue, who pass MacArthur and Patton and Thayer, might reasonably expect that their careers will not take them to four or five stars and the command of hundreds of thousands of soldiers engaged in a global war. But any cadet might one day find himself or herself faced with an unforeseen crisis, with the chance to do something courageous. The plaques and the statues tell the cadets what feats others have accomplished in their crucial hours. Like Bucha's Medal of Honor, which hangs in a shadow box outside the entrance to Crandall Pool, they remind cadets of what they may be called upon to do.

The Superintendent's office sits high in the stone tower of Taylor Hall, at the top of a staircase of polished stone. Embedded in the walls are trophy cannons from the Mexican War. There is a large foyer

outside the Superintendent's office: The floors gleam, the leather furniture is sedate, the lighting muted, and the whole area feels more
like a cathedral than an administration building.

Just outside the big oak doors sits a colorful display of photographs showing a wide look of cadet life. In one photo, taken at a football game at Michie Stadium, the Superintendent, Lieutenant
General Daniel W. Christman, USMA '65, leaps into the air to
bounce off the chest of A-Man, a caped superhero in black and gold,
Army's unofficial mascot.

The cadets love Christman for this exuberance.

"He's great at the football games," Cadet Jacque Messel says.
"Running around with A-Man."

Christman's energy is on display as he strides into the Cadet Mess
amid the crowd of gray pouring in for lunch. He is a big man, six two,
two hundred pounds, broad in the chest and shoulders. He has white
hair and a healthy, florid complexion; he is almost always smiling.

"Hello, hello," he greets the cadets, many of whom he knows by
name. They move to get out of his way, but they aren't scattering.
They move because he's the Supe, a three-star general, and military
courtesy demands that they make way. But they don't move far; they
linger in his line of sight, meet his eyes, smile when they greet him.
They like being around him. He gives them hearty, two-handed handshakes, pats them on the shoulder, leans close when they talk to him.

Christman has just come from three long hours trapped in his
office: meetings with his staff and speechwriters, telephone calls and
letters, strategic planning, and community relations. He's come to the
Mess Hall specifically to congratulate the women's basketball team on
the weekend's victory over archrival Navy. Here among the troops,
Christman is like a kid let out of school.

Around him, hundreds of tables are crammed in tight quarters,
and there is a din of noise as the cadets jostle each other, shout to
friends, give commands to plebes. Christman attracts attention, and
the cadets at nearby tables crane their necks to catch a glimpse. In
the short wing of the old mess hall, which dates from the twenties
(and where Cadet Christman ate many of his meals) he finds the

women's basketball tables. He greets the firsties, who have played their last game against Navy. The seniors introduce the underclass players to him.

"Don't get up," he tells the underclass cadets, who push their chairs back at his approach. The team captains stand deferentially, happy to be singled out for the honor of a visit. When the cadet adjutant calls the corps to attention, Christman snaps to, straight as any plebe, until the command, "Take seats."

"I'm going to say hello to the hockey team," Christman tells his aide as he heads to another wing of the Mess Hall.

Under a twenty-foot stained-glass window depicting Washington, Christman greets a headwaiter in the same way he does his favorite athletes. He finds the hockey team in the corner, shakes hands with the cadets at the head of the table and chats them up amid the deafening roar of the Mess Hall. They are all smiles and good cheer because the leader has come to see them, has walked over on his way home to say a few words to the troops.

"It takes a great deal of talent to be a good Superintendent," Lieutenant Colonel Jeff Weart, of the Commandant's staff, says. "There aren't many guys out there who have what it takes. You've got to have the intellectual ability and the academic credentials to head up a major college. Christman has that."

Daniel W. Christman was first in his class of 1965, a singular honor the Academy mentions in every press release and official introduction.

"The guy earned a law degree in his spare time while he was working at the Pentagon," Weart says, shaking his head in disbelief, not at Christman's considerable intellectual ability, but at the fact that someone could find free time while working at the Pentagon.

In addition to the Doctor of Jurisprudence from George Washington University, Christman holds two master's degrees from Princeton (in Systems Engineering and Public Administration). He can hold his own with any college president.

"A Supe also has to have all the Army credentials. He has to have done the general thing."

LEFT: R-Day: the Class of 2002 prepares for their swearing-in ceremony. *(Ted Spiegel/USMA Admissions)*

ABOVE: Former Army linguist Ben Steadman on the assault course. *(Photo by author)*

LEFT: Squad leader Shannon Stein on the assault course. *(Photo by author)*

ABOVE: Tom Lamb, who spent a year at the University of Portland,
provides covering fire on the assault course.   *(Photo by author)*

LEFT: Shannon Stein's squad at Lake Frederick:

Standing, left to right: Bob Friesema, Pete Haglin, Tom Lamb.

Kneeling, left to right: Barry Degrazio, Clint Knox, Jacque Messel, Marat Daveltshin, Pete Lisowski.

Sitting: Ben Steadman.

Not pictured: Omar Bilal.
(Photo by author)

LEFT: Shannon Stein on the march back from Lake Frederick at the end of Beast.
(Photo by author)

ABOVE: "Most of the sixteen-plus miles are done on dirt roads that wind through the woods. It is humid under the trees, and the new cadets walk in silence, one file on each side of the trail." *(Photo by author)*

LEFT: Bob Friesema using his break time on the road march to study plebe knowledge.
(*Photo by author*)

RIGHT: A new cadet plays the bagpipes for his classmates on the road march from Lake Frederick.
(*Photo by author*)

LEFT: Shannon Stein near the end of the road march from Lake Frederick and the end of her duties as a Beast squad leader. The Beast cadre is "eager to turn over their charges, let them be someone else's problem."
(*Photo by author*)

LEFT: Sergeant First Class Donald Mercier and Sergeant First Class Tim Bingham near the end of the road march from Lake Frederick. Regular Army NCO's are assigned to teach the cadets NCO duties.
(Photo by author)

RIGHT: Kevin Bradley, USMA 1999 The gold stars denote the top 5 percent in general order of merit. Bradley graduated number 20 in his class of 937.
(Academy photo)

ABOVE: Major Rob Olson, USMA 1987, (left) and Captain Brian Turner, USMA 1991. Olson is the senior Tactical Officer for Alpha Company during Beast; Turner is the associate Tac, an understudy preparing to assume the role.
(Photo by author)

LEFT: Bob Friesema, USMA 2002
"You have to know that tough times are necessary, and it'll be better for you in the long run." *(Academy photo)*

LEFT: Jacque Messel with father Robert, USMA 1968, and mother Jerri. *(Photo courtesy Jacque Messel)*

RIGHT: Cadet Pete Haglin, USMA 2002
"West Point is, for him, something to get through on his way to the 'real' Army he knows from his father's stories, from his experience growing up an 'Army brat' on posts all over the world." *(Photo by author)*

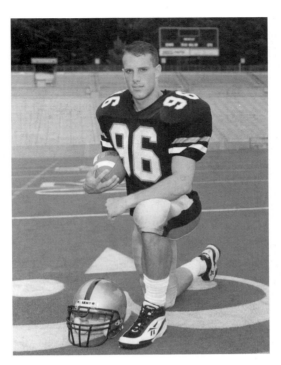

Grady Jett, USMA 2000
(Photo courtesy Army Sports Office)

Shannon Stein (center) and some of the members of her
Beast squad.   *(Photo courtesy Shannon Stein)*

Shannon Stein and her grandfather, John Baker Welsh. Welsh, a World War II vet, was a First Lieutenant in the Army Air Corps. *(Photo courtesy Shannon Stein)*

John Norton, USMA 1941 (second from left) in Normandy about D+5. Nearly sixty years before Beast Barracks for the Class of 2002, Norton believed that developing junior leaders was the key to success on the battlefield. "You watch 'em, you coach 'em, you trust 'em." *(Photo courtesy John Norton)*

Greg Stitt and Sarah Hatton. The classmates were married the day after graduation in May 2000. Stitt, a platoon sergeant during Beast, was widely admired by the new cadets, who worked hard because they didn't want to let him down.
(Photo courtesy Greg Stitt)

Alisha Bryan (right) and a friend.   *(Photo courtesy Alisha Bryan)*

Alisha and companymates after Christmas dinner in the Cadet Mess.
(*Photo courtesy Alisha Bryan*)

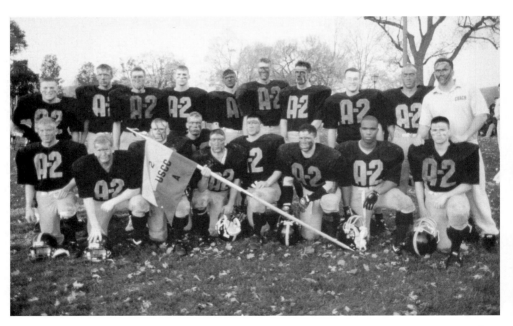

Intramural football team, company A-2. Colonel Maureen LaBoeuf, head of the Department of Physical Education, says, "We have more athletic opportunities for four thousand cadets than Ohio State does for its forty thousand students. The physical program at West Point is the most rigorous, the best program in the country."
(*USMA Admissions/Ted Spiegel*)

A three-person cadet room.
(*USMA Admissions/Ted Spiegel*)

Kevin Bradley, center, and Sandhurst teammates before the competition on a Saturday morning in spring.
(*Photo by author*)

Kevin Bradley as a human ladder in the Sandhurst competition. Note the ubiquitous "Ranger" decoration. *(Photo by author)*

Cadet Murphy Caine leads the cheering section for Sandhurst. *(Photo by author)*

The view north across the Plain to Battle Monument and the Hudson River. Greg Stitt climbed this shaft in a "spirit mission." (USMA Admissions/Ted Spiegel)

Captain Carol Anderson, Department of Physical Education, the first woman to teach boxing at West Point. (Photos courtesy Phil Kamrass)

ABOVE: The Class of 1999
prepares to review
the rest of the Corps
in the graduation parade.
(Photo by author)

ABOVE: Dave and Madge Bradley with Kevin just
after the Graduation Parade.   *(Photo by author)*

RIGHT: Brand new Second Lieutenant Kevin
Bradley immediately after the hat toss.
Next stop, Fort Knox, Kentucky
and—within the year—Kosovo.
(Photo by author)

ABOVE: Graduation and the hat toss. "Instantly the hats sail into the bright blue." (*Photo by author*)

LEFT: Lieutenant General Daniel W. Christman, West Point's Superintendent. (*Photo courtesy Army Public Affairs*)

RIGHT: The Superintendent as football fan. (*Photo courtesy Army Public Affairs*)

LEFT: Brigadier General John P. Abizaid, USMA 1973. As Commandant, Abizaid is responsible for the military development of the cadets.

"Leadership is the most dynamic aspect of combat power . . . . No technology can give you the advantage that good old-fashioned leadership can give you. You can have spy planes overhead and all kinds of information downloaded to the G2 [the intelligence section]. If the lieutenant and sergeant on the ground don't do their duty, we will fail." (*Photo courtesy Army Public Affairs*)

RIGHT: Colonel Maureen LeBoeuf. Her job is head of the Department of Physical Education; her title, a legacy of the nineteenth century, is Master of the Sword.

"Athleticism is a secular religion at West Point; Maureen LeBoeuf is the high priestess. Her computer screen saver says, 'Surround yourself with good people, delegate authority, give credit and try to stay out of the way.' " (*U. S. Army photo*)

In order to be respected among his colleagues who wear stars and run the Army, in order to be a voice among the other major commands, whoever occupies the big office in Taylor Hall must have commanded at the highest levels. Christman led an engineer company in Vietnam, a battalion in Germany, and several directorates in Washington before becoming the commanding general of Fort Leonard Wood and the U.S. Army Engineer School, responsible for training all the Army's combat engineers.

"And even all that isn't good enough," Weart says. "The guy has to connect with the cadets if this place is going to run well. And not everyone can do that."

Weart mentions a few names. Howard Graves, Christman's predecessor, "acted like he was afraid of cadets."

Dave Palmer, who was Superintendent from 1987–1992, was also incapable of connecting with the cadets, staff, or faculty.

"No personality," Weart says. "Christman has it all. The cadets love him."

Christman needs all this influence because of the paradox of leadership at his level: He has very little direct control over what happens at West Point. He leads by keeping focused on the mission, by providing opportunities for his subordinate leaders to make things happen, by setting the course. This hands-off approach takes imagination and energy. It also takes more than one man's share of self-confidence, and no small amount of courage.

Outside the administration building, Thayer Road is awash in gray uniforms as hundreds of cadets head back from class. General Christman steps outside, then makes a little clicking noise. He is doing exactly what the most junior instructors do: he tries not to be outside when the street is full of cadets, because they all salute and they all want him to salute back, one at a time.

An academy van is parked by the curb in front of Taylor Hall, and two NCOs stand by the open door. Christman greets them by first name; they call him "general." One of the NCOs gets the two-handed shake.

The van crawls slowly through the area. Most traffic is banned here, so the cadets walk in the street. They also salute—there is a license plate with three stars on the front—and Christman waves back, like a popular mayor. He says aloud the names and sports of a half dozen cadets he sees, and every cadet he names is a Corps Squad athlete.

The only criticism of Christman commonly voiced among the faculty and staff is that he focuses too much on Army sports and Corps Squad athletes.

His visit to the Mess Hall earlier that day was for the express purpose of congratulating the women's basketball team. On the way through the crowd, Christman stopped to chat with two huge cadets, both football players. After greeting the basketball players he waded through the crowd to a far corner of the huge room to say hello to the hockey team.

Asked if the Superintendent knows any cadets who aren't athletes, his aide, Major John Moellering, hesitates before answering. It could be that the question has not occurred to him before.

"Well, he knows the chain of command," Moellering says. "The brigade staff. And the Honor Committee."

Much of this criticism of Christman comes from the academic side of the house. According to one professor, at every briefing of the staff and faculty, the Superintendent trots out various Corps Squad captains and sports stars. "But I couldn't tell you what the top-ranked academic cadet looks like, or who the scholarship winners are."

In a meeting with the presidents of West Point's alumni organizations, who gathered at the Academy in August, Christman spent nearly forty minutes of his hour talking about sports facilities. To dramatize the need for new locker rooms in the football stadium, he used orange traffic cones, set up on stage, to show the size of the shower room used by the football team.

Another of his slides showed Army's record against Navy for the last few years: the cadets performed below .500 across all Corps Squad contests.

"Only in one year since 1970 has Army won more than half its contests with Navy. I'm here to field competitive teams on the fields of friendly strife," Christman told the audience of alumni.

At one point in the briefing Christman put up a simple, two-axis graph showing how West Point's sports facilities have declined in quality since he was a cadet athlete in the early sixties. One line on the graph shows how the facilities of civilian universities have improved in that same period. He offered no explanation as to the source of the information. Then he turned to the audience and said the graph is "subjective"; he made it up to illustrate his point.

Christman then showed the alumni photographs of the decrepit rifle and pistol range, the medieval locker rooms the track team uses, the overcrowded weight room. He compared these to photographs of the same facilities at Navy and Air Force, which are sleek and modern.

Christman insists that the facilities directly affect recruiting because they have a negative impact on candidates who visit West Point.

"Does this look like it belongs to a school that's serious about athletics?" he asked the audience as slide after slide went by. "Modernization of our sports facilities is a nuclear arms race with civilian universities."

That kind of statement makes many people at West Point shake their heads in amazement at Christman's priorities. But the Superintendent knew his audience, and he knew his opportunities. There would be no modernization of sports facilities unless the alumni coughed up the money. Near the end of his talk, he held up a photocopy of an article that had appeared in *Sports Illustrated* magazine, a piece entitled "Seventy-Six Things Right about Sports."

He paused, looked down at the page in his hand, then up at his audience.

"Army versus Navy in anything."

The van drops him at the Officers' Club, and the Supe hustles down the steps to the barbershop. He greets the young woman barber as he does everyone: big smile, eye contact. "How are you? Good to see you."

"Be careful you don't break your comb in that really thick stuff on top," he says, patting his nearly bald head. The barber smiles, clearly charmed.

As he sits in the chair, Christman considers John Norton's prescription for growing leaders: You watch 'em, you coach 'em, you trust 'em.

"I would say you have to turn that around," Christman says as the barber snips at his thin hair. "And put trust at the top. The leader has got to engender a feeling of trust in his subordinates; they have got to trust him or her. That's how you build a cohesive unit—trust between the leader and the led. It comes from the confidence the subordinates have in the competence and character of the leader, in his ethics, in his integrity. The troops pick up on all that stuff right away, and they can spot a fake instantly."

Like Abizaid, the Commandant, Christman believes that people have to have the opportunity to run things on their own. In his own career, he learned more when he was allowed to "run with the ball." Now, as Superintendent, letting his subordinates run with the ball means he has to be willing to support them when they fail.

Christman had a chance to affect the command relationship Army-wide when he served with the Army's Training and Doctrine Command, the agency that spells out how training is to be conducted. He was concerned about commanders being in too deep, about two-star generals hammering subordinates four levels down about the most minute details of training and maintenance.

"They took a good idea—check on things—and carried it to a ridiculous extreme. That will squash initiative. It's antithetical to what we're trying to accomplish. We want to have comfort with ambiguity. The more you resist that, the more inconsistent we are with the current environment for junior officers."

"When I was a firstie on the Beast detail, I was in charge of the R-Day ceremony on Trophy Point," he says, describing the swearing-in ceremony conducted for the new cadets. In rehearsal with the cadre, Christman says, "I bollixed things up. I was trying to move people; they were all in the wrong positions; it was a mess."

"The officer in charge was a Major Rhyne. He let me go long enough so that I could see what I had done. Then he said, 'Let me step in.' And what he meant was 'let me step in and help you without making you feel like an ignoramus.' And he did a couple of small things and bailed me out. He let me go far enough to learn the lesson. What I admired most was how he made the corrections. Very mature, nonthreatening."

"Later on I had a similar experience introducing the Superintendent, General Lampert. I mispronounced his name, and he came up to me and said—very quietly—'Dan, just remember, it's with a *p*, not a *b*.'"

Christman smiles at the memories, but these are more than just good stories, they are part of what has shaped his near fanaticism about the interaction between leader and led. If he beats a drum other than Army sports, it is this: leaders respect their subordinates, and demanding does not mean demeaning. Christman and the recent Commandants have forced an evolution in the Corps of Cadets. It is his handiwork that guarantees that plebes are stressed without personal, demeaning attacks.

When his haircut is finished, Christman thanks the barber and heads out into the cold. The sunshine has warmed some of the sidewalks, melting ice and leaving dark streams of water on the sidewalks. He crosses the cadet area and enters the front doors of Quarters 100.

Christman strides into a coatroom as large as some of the living rooms in on-post junior officers quarters. There are racks of hangers; sports equipment fills the corners; a "Beat Navy" pennant is tossed on a shelf top. It looks like the mudroom of a house full of active teenagers.

Christman greets the household staff and asks that lunch be served in a sunny breakfast room just off the walled garden. Outside the big windows, the upper floors of the cadet gymnasium are visible above the brick wall. One wing of MacArthur Barracks looms over the southern end of the garden.

"We have to keep in mind that we're not just preparing these cadets to be lieutenants. They're going to be stepping up to the plate in 2024 or so to run the whole army. How are they going to handle terrorism, anti-narcotics, the nuclear threat, as well as be able to fight in regional conflicts?"

The sandwiches are tuna fish and lettuce on wheat bread; they come out on glass plates wrapped in plastic. There are Diet Cokes, and ice in glasses from the Class of '99 Ring Weekend. The class crest is engraved on the side of each tumbler.

"Our world, the world we graduated into, was bipolar," Christman says. "One of my biggest concerns is that we turn out agile thinkers, because the world isn't so simple anymore. We have to emphasize creativity and nonlinear thinking."

He swirls his Coke in the glass. The Academy motto, Duty, Honor, Country, visible on the side, makes it the perfect prop.

"And we have to do that while holding to these classic values."

Christman spends less than ten minutes at lunch. He thanks the staff, then heads to the front of the big house. The Lee Room, just off the entrance hall, is set with thirty-some folding chairs, and the room is big enough that it doesn't seem crowded. There are plates of cookies on the low tables. A large oil painting of Robert E. Lee in his U.S. Army uniform hangs above a fireplace with a green marble mantel. This is an unfamiliar Lee, dark hair and mustache, no beard, Union-blue tunic. Lee, Superintendent in the 1850s, lived in this house and used this front parlor as an office and reception area. His desk and a clock are in one corner.

Christman has invited the juniors just back from the fall semester's exchange program in for a chat. Christman sits on a bench beneath a wide window as the cadets file in; the back seats fill up first. These are some of the top-performing cadets of the class of 2000, who were sent out for a semester to represent West Point and learn from other academies. They went to Navy, to Air Force, to the Coast

Guard Academy in New London, Connecticut, to St. Cyr in France. Christman passes around a plate of cookies, then jokes with the cadets about the privileges they enjoyed at their host schools.

The cadets are not afraid to either praise or criticize West Point. One young woman talks about how being away from West Point made her appreciate that there are reasons why the Academy does things in a certain way.

"It helped me fight off cynicism," she says of her semester at Air Force.

The conversation ranges widely, from the professional and technical training of Coast Guard cadets, to hazing at Air Force, to how many washing machines are available to Navy's midshipmen and how much they cost. Christman takes it all in, writes some notes, smiles and jokes.

Brigadier General Fletcher Lamkin, the Dean of the Academic Board, is in the room; the cadets address many of their questions to him: about late lights, about the after-hours availability of labs and academic buildings. One cadet draws laughs from his classmates and the Supe when he refers to the Dean as "General Fletcher."

"Sure, you can call me by my first name," the Dean jokes.

The range of the discussion mirrors Christman's job. He must be concerned with everything from the size of the shower room used by the football players to the demographics of the nation from which West Point will draw its classes for the next few decades. He thinks about the impact of technology on the Army and how best to prepare leaders who can manage information. He will be held accountable for the performance of thousands of lieutenants commissioned during his watch, as they pull duty at roadblocks in Bosnia, safe havens in Kosovo, and maneuvers in the California desert. He must consider how the Army retains officers and how West Point treats its plebes. He knows the record of the women's basketball team and who is being considered for the captaincy of the men's hockey team.

The job could be consuming, and Christman goes about it with an energy that might lead some people to believe it is all he thinks about. But he has some perspective, and a sense of humor.

At the end of an enthusiastic introduction to a group of alumni, the president of the Association of Graduates spoke of how much Dan Christman loves West Point. Christman rose and told a story on the AOG president.

"We were at a meeting in Texas, and Jack [Hammack, '49] gave me that kind of introduction. He ended by saying, 'And I know that every night, just before he goes to sleep, Dan Christman is thinking about what's best for West Point.'

"And my wife, Susan, who was sitting up front, said, 'Well, not every night, Jack.' "

EVERY CADET AN ATHLETE

Spring doesn't arrive at West Point until April, and even then its hold is shaky. The air can still be cold, but on sunny days, visitors can see the light green tint in the trees creeping up the side of Storm King Mountain, which looms in the distance behind the cadet gym. A plaque outside the entrance to Arvin Gymnasium notes that the facility is named for Carl Robert Arvin, USMA 1965. Captain of the varsity wrestling team, two-time All-American, and first captain of the Corps of Cadets. Captain Arvin was killed in action in October 1967 in Vietnam. He was twenty-four years old.

Inside the gym, a line of nervous cadets waits outside the door to the third-floor wrestling room. They wear camouflage BDU shirts over PT shirts and shorts. They are in their stocking feet. As they wait to be tested, they nervously chew on protective mouthpieces, bounce back and forth on the balls of their feet, laugh nervously, and study the grading charts lined up on one wall.

The cadets are all third class, yearlings. This morning's class is the culmination of hand-to-hand combat instruction: nineteen lessons for the men, nearly forty for the women. Like everything else at West Point, the standards are spelled out exactly: in this case on neatly lettered boards, although the grading here is more subjective than usual. The cadets in line are being graded on how well they protect themselves from attack, and how aggressive they are.

"Couldn't we all just get along?" one of them jokes.

The wrestling room has no windows, and only one light is on, so the sixty-foot square room is quite dark. Add to that the heavy-metal music pounding from overhead speakers at ear-splitting levels. Add to that a maze of wrestling mats turned on edge so that they reach from floor to ceiling, folded into twists and turns like alleys and short hallways.

A Cadet Jones is the first called into the room. Medium build, about 160, with dark hair and olive skin. When he enters, the first thing Jones sees is a dark figure, its arms, legs, and chest girded in pads, face covered with a wire mask. The room is too dark to see anything but the outline of this person. The music is loud enough to rock dental work.

"Hey, hey, hey," the dark figure says, approaching in a crouch.

Jones is not allowed to attack; he must wait to be attacked.

"There's a fight, you gotta break it up," the figure screams above the noise. Jones turns to the right, and, sure enough, two more shadowy figures are pushing and shoving each other, screaming and cursing.

Jones calls, "Hey, hey! Break it up!"

Suddenly one of the fighters whirls and attacks Jones with a baton (it's rubber), which means he can react. He sidesteps the blow, moves in, and delivers two or three quick punches to the face-mask of the attacker, then grabs an arm and throws the assailant to the floor, finishing him off with a flurry of kicks and punches.

An instructor tells Jones to move on. Just ahead is another padded figure, this one much shorter. The figure is quiet until Jones gets close.

"Hey buddy, hey buddy," she repeats, over and over. She's annoying enough to want to hit, but she hasn't struck, so Jones can do nothing. As he turns left into a narrow alley between the wrestling-mat walls, the short figure follows Jones, badgering him, jumping up and down, touching his elbow. Just ahead, another padded figure comes out of the gloom, this one with a bigger stick. Maybe even a bat.

But no one has said stop, and the annoying little bugger is still at Jones's elbow. "C'mon, c'mon," the figure in the alley screams, brandishing the big stick. "You want a piece of this? You want a piece of this?"

Jones steps to his right, hugging the wall for protection. He glances back over his shoulder at the bouncing follower, jumping up and down like a hopped-up troll.

The cadet moves to the shadow with the stick—what else is there to do?—when *BAM!* Something big crashes into him from the side, from a small opening between the wrestling mats. This one is a big— six feet, 180 pounds—and he pushes Jones up against the wall, gloved hands reaching for the throat. Jones swings wildly, staggers, then gains his footing, fights back, swings out from under the two-handed grip and delivers three hard kicks to the big man, who falls to the floor. Jones get in a few more shots for good measure.

Down the alley. The screamer has disappeared. Turn the corner; another figure beckons in the distance, jumping up and down. As Jones moves out, someone else runs up behind, plows into him, wraps him in a bear hug. Jones sidesteps, ducks, goes down on one knee and twists free. Spinning, he throws hard, straight punches as the attacker rolls into a ball.

Now Jones is on his feet, blowing hard, face red with exertion. An instructor in gray appears, points to a far corner of the room, where two more padded figures grapple. Jones has taken only a single step when one charges him, a full run from twenty feet away, arms upraised, screaming. Jones goes stiff for a second, then dives forward toward the attacker, who must jump at the last second to avoid being bowled over. Jones scrambles crablike, follows the rolling attacker, kicks and punches and kicks and punches with real fury now, until the gray shirt comes up and taps him on the shoulder.

This is City Streets, where cadets practice what they've learned about fighting hand-to-hand.

One instructor is fond of telling cadets that the reason they take boxing is, "So that you know you ain't gonna die if someone punches you in the face." It is also about aggressiveness; and there is always the chance that cadets may have to use these skills some day.

Major Jennifer Caruso, who might be an inch over five feet tall, tells the story of a woman cadet on Spring Break in Jamaica. "She was out on the dance floor with a bunch of friends, and this guy starts coming onto her, grinding her, you know? So she says to the guy, 'Back off!' "

Caruso does a fair imitation of how loud a woman would have to yell such a thing on a crowded dance floor. She stands, muscles tensed, feet apart, balanced; she is ready to attack. The tension, the coiled potential energy, comes off her in waves.

"But the guy doesn't get the hint, so he starts crowding her, then he grabs her. So—*wham!* She pushes his arms away and smacks him in the chest."

"Then she hit him again and the guy went flying. She said the whole dance floor moved over and cleared a space around them," Caruso continues. "The guy got up real fast and got out of there."

Caruso nods, proud of what her student learned.

Instruction in what West Point calls "combatives," the umbrella term for anything involving fighting, is not just about getting through Spring Break or surviving a mugging. The Academy says its young initiates must do things outside their comfort zone: combatives qualifies. For the men, it is boxing and, during third class year, wrestling. There is an elective called "grappling," a combination of collegiate style wrestling and fighting techniques borrowed from other disciplines. There is even an advanced hand-to-hand-combat course, another elective cadets can take during the "life sports" section of their education, in place of golf, for instance.

The course the plebe women take is "Close Quarters Combat." Jacque Messel's class meets at the relatively civilized hour of 8:30 A.M. The fifteen plebe women in BDU shirts over PT uniform are more talkative than the men gathering on the first floor for boxing. When they enter the room, they remove their shoes and line them up in neat ranks along the edge of the mat, like a ghost formation. Then they start practicing the falls and the blocking movements they'll use against attackers.

The big wrestling room, sixty feet square, is lit by harsh fluorescent lights, which create a low hum in the background. The floor is covered with football-style blocking dummies and heavy bags. One of the instructors (they're all women in here) is pregnant and wears a huge gray DPE shirt that billows out when she walks.

Most of these cadet women were athletes in high school; for many of them the standard physical education program was a joke. The most difficult part of the PE program in most American high schools is the angst of insecure teenagers worrying over whether or not they should undress in front of their peers. Not so at West Point. Here one had better master things quickly.

The instructors pile the bags on the floor, and within a minute or two of class starting the women are diving over into forward rolls. The ones who master it quickly move on. The ones who have difficulty, and Messel is one of those, stay behind and get more attention from the instructors.

Captain Carol Anderson, the primary instructor, watches from the side. She is the same woman who led the Commandant's staff in swimming.

"I ask all the women I teach," Anderson says, "if they've ever been in a fight. I say, 'Not a tug-of-war with your little brother over a CD. I'm talking about a fight where someone has tried to hurt you, where you have tried to hurt someone else.' Not surprisingly, 99.9

percent have never been. If someone tries to attack them, beat them up, rape them, they'll at least have learned something they can do to protect themselves."

Anderson dominates the room. She is five ten, with the thin, muscular build of a sprinter or basketball player; a short ponytail; little wire-rimmed glasses; commanding voice. On her black shorts she wears badges for top scores on both the Army's physical fitness test and West Point's dreaded indoor obstacle course, nemesis of cadets. Instructors in DPE don't have to take the test. Some, she says, take it as a "gut check."

Near Anderson, two women practice a move to block a direct punch, then counter by pulling the attacker's head down into an upcoming knee. The women pull their punches and one of them says, "I don't want to mess up your hair."

"A lot of them think they don't need this," Anderson says. "And I get more attitude from the women than from the men."

Anderson gathers the women in close for a demonstration. Her moves are graceful, deliberate, and rapid as a snake strike. She smiles as she goes through the choreography, talking about oblique movements and kinesthetics, mixing a little plain English.

"If you do this," she says, dropping her guard, "You're gonna get whacked.

"When you do a heel strike," she says, shooting her arm out and at the nose of a cadet who stands just out of reach, "I want you to strike *through*, which means that I'm not trying to strike the nose, I'm trying to reach through to the back of the head."

Around her, a couple of cadets practice the move silently.

"When you do the groin strike, hit hard," she continues, swinging her arm down sharply at the groin of a woman standing close behind her. Her large hand is open.

"Then pull and twist," she says, demonstrating a move quick enough to start a cold lawnmower. When she says, "That's one of my favorites," a couple of cadets chuckle.

Anderson tells a cadet with a club to come at her. It is the classic move, a staple of every movie about basic training, where the instructor invites a trainee to attack. The cadet approaches, and Anderson delivers a rapid combination: heel strike to the face, knee to groin, knee to head, and—a *coup de grace* as the attacker goes to the mat—a groin pull.

Later, as the cadets practice, Anderson watches a woman punch and kick a much bigger partner. For many of the cadets, this is about passing the course. For Anderson, it is something else.

"Maybe three out of ten are really trying, the others do the minimum they can to get a good grade," she says as she watches.

Messel and her partner move at less than half speed.

"But they're a lot more sincere in Self-Defense II, where we talk about rape attacks.

"I tell the cadets I've been the target of three attempted rapes since I've been in the army. And this wasn't out in the desert with Iraqi soldiers. These were soldiers or husbands of soldiers. And I'm not the typical rape victim. I'm five ten, 150 pounds. Usually victims are small women, so if it can happen to me it can happen to anyone," she says. "Eighty percent of the girls get it then."

Anderson didn't learn tough in a classroom, or in any Army training film. She grew up in Compton, California, which she says is "much worse than southcentral L.A. in terms of gang-bangers."

Anderson lived with her grandmother in Denver for a while, but her mother got sick and her father asked her to return to Los Angeles.

"I wasn't there a week when I knew I wasn't going to last at Compton High.

"They were doing stuff in eleventh grade that I had done in seventh grade. You mostly just went to class and hung out and went to the next class and hung out."

Anderson's brother and sister were involved with gangs, and in her neighborhood young women either wound up pregnant, in jail, or dead. She wanted no part of that. With money saved from a summer job, she bought a plane ticket to Denver, flew out on her own and called her grandmother from the airport. But her plan didn't work, and she soon found herself back in Los Angeles. Soon after she took the high school equivalency test, got her diploma a year early, and enlisted in the Army as a way out of Compton.

As an enlisted soldier, Anderson was a supply specialist. She went to Officer Candidate School at Fort Benning, Georgia, in 1989 and was commissioned in the Corps of Engineers. In 1991 she went to Desert Storm with a combat engineer (heavy) platoon; she took bulldozers, road-graders, front-end loaders and their soldier/operators to war. Her outfit, designed to build supply roads and airfields, was not equipped for combat. It relied, for protection, on other units.

"We were attached to this battalion of combat engineers [a unit equipped to operate close to the front]. I had this raggedy old CUCV [a civilian sport utility vehicle painted camouflage], and they had humvees and nice equipment. We're just following them along in the desert. I didn't even get a map, much less a 'slugger' [a hand-held global positioning system]. When we stopped everybody went to sleep: all the soldiers, all the squad leaders, my platoon sergeant. I got called to battalion headquarters. While I was there, the commander says, 'We're leaving. You need to meet up with this other battalion.' He was on the radio, talking on two nets, taking and giving orders. He gave me about a minute or two of his time. I didn't even have a compass, just a map I had taken from someone else. He said, 'We're going this way and you're going that way.' It was like, 'Take charge. See you later, lieutenant.' He abandoned us."

Before Anderson could even get back to her troops, the engineer unit was rolling. She woke her platoon sergeant and told him they were being abandoned. "You can't let them leave us out here!" he screamed at her.

"I said, 'Too late, it's happening.' He told me I was the dumbest second lieutenant he'd ever seen."

Anderson smiles telling the story, but leaves no doubt that it was a frightening night. She didn't know the enemy situation; she wasn't sure where the friendly units were. She wasn't even sure which direction to go. She and her platoon sergeant knew they were in Iraq, but they didn't know if they were going to run into Iraqi or American units.

"There's nothing out in there to navigate with, no terrain features.

Just sand in front, sand behind, sand to the left and right. My platoon sergeant happened to have a compass—we didn't even get issued stuff like that!"

"He and I disagreed on which way to go. He wanted to go one way and I wanted to go the exact opposite direction. Finally, he said, 'Whatever. We'll know how wrong you were when we get killed.' "

Anderson and the platoon sergeant decided not to tell the soldiers their predicament. They acted like they knew what they were doing. The only one privy to the disagreement between Anderson and the platoon sergeant was her driver. They began moving, and Anderson finally spotted a vehicle, tiny in the vast flat expanse of sand.

"I couldn't tell if it was a friendly or not, but there wasn't much I could do about it at that point. He started buzzing towards us, and I see it's a humvee. He comes pulling up and this lieutenant jumps out and we're both so relieved that we hug each other. His mission was to find us, but he only had another fifteen minutes or so to wait and then he was gone."

"We'd have been left there in the middle of the desert. To this day I couldn't tell you how I found that humvee. My platoon sergeant and I never got along before that; this changed it all. He apologized and told me he was glad I didn't listen [to him]. We navigated by his compass and by the grace of God."

West Point puts officers like Anderson close to cadets so that the cadets can learn from these experiences. But Anderson says, "I don't tell the cadets [that story]. I still feel a little like that dumb second lieutenant."

This image is hard to fit with the woman in the wrestling room, who is the very picture of self-confidence. In fact, Anderson is the first woman to teach boxing at West Point. She shrugs, gives a no-big-deal look.

"It's just a class I teach. In fact, it's one of the easiest ones I teach. It's all about movement."

Perhaps, she allows, boxing comes easy to her because she was a fighter back in Compton, California.

"I got suspended for fighting probably three times a year in high school. And that's just for the times I got caught. I got expelled from grade school. That was just how it was in Compton. We got picked on a lot. Had these raggedy old hand-me-down clothes and our shoes were from the department store and not the nice kind. Kids would ridicule us.

Anderson says she never started a fight, but she never walked away from one either. Not surprisingly, a lot of what she has to tell cadets has to do with tenacity and perseverance. You can accomplish a lot, she tells them, if you're willing to work hard. But there is an undercurrent of discontent when she talks about cadets.

"I expected a lot of motivated young people excited about becoming Army officers. I don't see that. You do have a percentage of 'high speed-low drag' cadets, then you have the great majority who are trying to get by. And another small percentage who aren't doing anything and are still here.

"Some cadets are here after doing things that would get a soldier thrown out of the Army. I've busted my butt to become an officer. For someone who's worked as hard as I have to get where I am, it pisses me off to see people scraping by and still here. There are lots of kids out there, just as smart, who would love to have this chance, and some of these kids are just cruising here."

It is the end of the period, and the plebes glance at the clock as they stomp into their shoes. Anderson gathers her clipboard as the cadets race off to the next requirement. "They know they're going to make it. The government brings you here and spends all this money, they want you to graduate. The message is to push people through."

The Department of Physical Education is the department cadets love to hate. It's just too easy to make fun of their tight uniforms, their big sports watches, the go-to-hell stance they adopt in the department photo hanging just inside the door to the gym.

"Nobody likes them," Bob Friesema says of DPE. "And I'm not just talking cadets. My math professor told us about this lunchtime basketball league the faculty has. Nothing gives him a bigger thrill than beating DPE."

The DPE instructors are drill-sergeant picky about uniforms, military courtesy, standards.

"In high school if you played to your ability you got a good grade," Friesema says. "Here, it's 'You meet the standard' or 'too bad.' You have to be an outstanding athlete to get an A."

Friesema's comments aren't the complaints of a teenage couch-potato. Although he had no experience boxing, he intimidated opponents just because his height gave him an advantage. At six four, he can outreach most opponents. He was a competitive runner and swimmer as a kid. Now he works out every day with some combination of running, lifting weights, pull-ups, push-ups, and sit-ups, for forty-five minutes, just to stay proficient and able to pass the tests.

West Point tells prospective cadets that the academy is a physical place. The admissions website specifies that candidates must have "above average strength, endurance and agility." A commercially published handbook for candidates shows a photograph of cadets playing a game of pick-up basketball. The caption reads, "If you don't like athletic competition, you will be a misfit at West Point."

The beginnings of this emphasis on athletics came from Douglas MacArthur's observation that young soldiers in World War I respected officers with athletic ability. When he became Superintendent after the war, in 1919, MacArthur decreed "every cadet an athlete." Participation in sports, either varsity, club, or intramural, has been mandatory for most of the twentieth century.

Athletics have gone from a good idea to a religion at West Point. And, like religion, it can have its excesses.

One woman who graduated in the late eighties says she was bulimic before she went to West Point, was bulimic her whole time as a cadet, and stayed that way until she became pregnant with her first child (she is now the healthy mother of four).

The pressures to conform to an ideal body type, already heavy on young women in America, are exaggerated at West Point. This woman, who won a varsity A in track, said that West Point did little or nothing to help women with eating disorders. She mentioned a classmate who was an All-American athlete, but who also had an eating disorder. She contends that coaches and officials ignored the signs because the woman was such a successful athlete. "They figured, 'If she's winning, she must be healthy.' "

Some cadet women are afflicted with "exercise bulimia", they eat, then exercise feverishly to work off those calories. This behavior is the easiest to disguise, since exercise, in West Point's culture, is always a good thing.

"I used to get up and run at 0515," the former bulimic says. "And there were a surprising number of people out there. And a good number of them were women."

But the presence of eating disorders is no longer a dark secret at West Point, if it ever was.

Colonel Maureen LeBoeuf's office in Arvin Gym is long, narrow, and windowless. There are neat chairs upholstered in blue and a small couch in front of a polished coffee table. On the walls hang mementos: a flag bearing the ivy leaf patch of the Fourth Division; a guidon with the winged wheel of the Transportation Corps and the unit designation of a helicopter company. A handsome, framed print shows Federal and Confederate cavalrymen clashing at the Battle of Brandy Station, Virginia in June 1863. The original was commissioned by LeBoeuf's class at the U.S. Army War College. At one end of the wall hang framed diplomas—a master's degree and a Ph.D.—both from the University of Georgia.

On this spring morning LeBoeuf is dressed in her "Class B" dress uniform of green slacks, light green shirt, and small tie. The black epaulets on her shoulders bear the silver eagles of a full colonel. Her dress coat, with its medals, insignia, and pilot's wings, hangs on a rack for a meeting later in the day. Her computer screensaver says, "Surround yourself with good people, delegate authority, give credit and try to stay out of the way."

On the coffee table in front of her are three binders; one of them is titled, "Eating Disorder Task Force."

"That's something new," she says. "Somebody briefed the Comm, told him this was a problem. The Comm told us to put together a task force to study the problem, see how widespread it is, and set up some support mechanism for cadets so they can get help, so we can identify the ones who need help.

"Eating disorders might be a little more common [among cadets] than in the general population," she says.

"Look at what we do. We bring in people who are very competitive, who are even prone to obsessive behavior. Then we put these young women in what is basically a man's uniform."

She touches the belt of her own uniform. "You put a young adolescent girl in a uniform with a belt and she better be pretty slim.

"And of course there's all the pressure from our culture. We associate fitness and thinness when really there isn't necessarily a connection."

The Eating Disorder Task Force includes representatives from the medical command, the dental command, the nutritionist, counselors from the Cadet Counseling Center. LeBoeuf represents DPE.

"It's amazing how a young woman will look at herself in the mirror and see only the bad things, while a man will look and see the good things. You'll see a guy with a big gut and a fat rear, but he's wearing spandex because he thinks he has great biceps," she laughs. "And he just looks in the mirror as he's working out and says, 'Man, what a great bicep.' Women look at themselves and only see what's wrong."

As part of her doctoral work, LeBoeuf studied the experiences of women in the classes of 1980 (the first with women), 1985, and 1990.

"At the end of all these standard questions I asked each woman: 'Did you have an eating disorder while you were at West Point?' There was this one woman I remember, she must have paused for thirty seconds before answering. She said she'd never even told her husband, but she had been anorexic."

"I knew I was dying,' she told me. 'I stood in formation and could feel the other anorexics around me. I used to look down on bulimics because they were weak—they ate."

"I asked her what got her out of it," LeBoeuf said. "She told me that she'd been dating this guy and he dumped her. A lot of women stop eating when something like that happens. She said, 'I started eating and haven't had a problem since.'"

But not everyone will be shaken out of the behavior. Women die from this disorder; cadets can be dismissed. One of the biggest obstacles officials face is getting cadets to admit they have a problem. Cadets are concerned that self-reporting will lead to separation. LeBoeuf mentions a woman who was having a dental exam. When the dentist took one look inside her mouth—she had a bunch of sores from vomiting—he told her, "I know what you're doing. I want to see you again in thirty days. If those sores haven't cleared up, I'm turning you in."

Fear—of exposure, embarrassment, even dismissal—weigh heavily on cadets' minds, but the physical culture is so powerful that it can be almost impossible to resist. Prejudice against people who are out-of-shape or overweight is the only acceptable discrimination in cadet culture. And, not surprisingly given the ratio of men and women in the corps (approximately ten to one), male attitudes dominate. LeBoeuf has made educating the men her mission.

"I talk to cadet men about eating disorders, and they don't really get it," LeBoeuf says. "One group of guys told me they'd say things to the plebe women like, 'You're not going to eat that dessert, are you? Or 'You look like you're putting on a few pounds.' And I tell them that's dangerous behavior. 'But we're only kidding.'"

She shakes her head. "That's their defense. I asked them, 'What do you think is going through the mind of this little seventeen- or eighteen-year-old, as she's sitting there surrounded by these older guys, all of them in great shape? Don't you think she's going to try to do whatever she thinks will please you?'"

The prevalence of eating disorders at West Point may be a perversion of the physical culture, but is isn't surprising. Athleticism is a secular religion at West Point, and Maureen LeBoeuf is the high priestess. The Academy uses every opportunity to tell cadets that, in the regular Army, soldiers don't care who was number one in history. They want a lieutenant who can finish the run and max the physical fitness test. They want a lieutenant they can brag about.

The Army emphasizes physical fitness and a healthy lifestyle for practical reasons. Soldiers perform better when they are physically fit; and that isn't just the tasks requiring strength and agility. Well-conditioned soldiers fight better, especially in the kind of round-the-clock operations that characterize modern combat. Their bodies handle stress better, and they are more resistant to sickness. The question is whether West Point has carried MacArthur's intention—every cadet an athlete—to an extreme.

Fifteen percent of a cadet's overall standing is determined by physical performance; 30 percent comes from military grades (which include leadership evaluations throughout the year). Fifty-five percent of a cadet's standing is based on academics. Yet a recent survey of cadet time use shows that cadets spend 22 percent of their "discretionary time" (that is, time not spent in class or at mandatory events) working out. It is difficult to draw a direct line between an extra hour of study and one's semester grade in physics; it is much easier to calculate the effect of an hour spent practicing the indoor obstacle course. And no one high-fives the cadet who gets the top grade in chemistry.

Cadets admit it is more acceptable (among cadets) to fail an academic test than it is to fail a physical test. The first is a test of intellectual effort; physical tests—and they are legion—are test of mettle, tests of character. They are, in the crude patois of the infantry, ways of "measuring dicks."

On another table in LeBoeuf's office is a copy of *Bugle Notes*, the "plebe bible." This one is dated 1949–1950. This may be a bit of irony, as it was the "old grads" who were most vehemently opposed to having a woman as the head of DPE. They launched an e-mail campaign, predicting the end of everything form physical fitness to the "warrior ethos."

LeBoeuf ignored most of what was said about her. "I just didn't want to waste the energy worrying about everything people might say." She got "incredible" support form the Superintendent, and from the Commandant at the time, Brigadier General (now Major General) Bob St. Onge.

In the spring of her first year as Master of the Sword, LeBoeuf went out on the speaking circuit for Founder's Day, when alumni associations around the world celebrate West Point's birthday with formal dinners. Tradition dictates that the oldest grad and the youngest grad in attendance give a speech. USMA leadership—most notably the Superintendent, the Dean, the Commandant, and various department heads—travel around the country to speak at these dinners. It is part celebration and part public relations, as the leadership updates the alumni on what's going on at West Point.

Many old grads see these dinners as a time to sharp-shoot high-ranking officers and complain about changes at West Point. This scenario became so predictable that Lieutenant General Christman sent out word to his hosts that he would not take questions from the podium at any of his appearances.

"I'll stand by the bar all night and talk if they want to listen," Christman said as he urged the other speakers to adopt the same strategy. "But I'm not standing around for some old grad to take pot-shots at me just for fun."

Maureen LeBoeuf stepped into this skirmish.

"Once they see that I don't have an eye in the middle of my forehead, once they hear that I have a sense of humor and that my knuckles don't drag on the ground, they get over most of their problems. I'm a lot harder to hate in person," she says, her face lighting with a smile. "Then they ask me questions like, 'How are the girls doing?' "

Because she can't help teasing the old grads, LeBoeuf used a routine that had been a favorite of a former Commandant. She produces a slide with several bar graphs: there is a pair labeled "push-ups," another for "sit-ups," another marked "Two-Mile Run," which shows finish times for that event. The scores for cadets in every other class, beginning with 1981 and ending with 1999, are compared to the scores for cadets from 1962.

The Class of '62 beat only one of the ten younger classes in push-ups (thirty-seven to thirty-six against the Class of 1983). All the younger classes beat '62 in sit-ups; the closest margin was eleven (1962 averaged sixty sit-ups in two minutes; 1981 averaged seventy-one. The Class of 1995 averaged eighty-eight). The older grads beat all the younger classes in the two-mile run.

"Then I uncover this," LeBoeuf says. The top of the slide is labeled "Women"; the slide compares the all-male class of 1962 with the women of the eighties and nineties. She laughs, enjoying the joke. She also agrees with the assessment, dogma among cadets, that women have to excel in order to be accepted.

"And the big place to do that, at least for new lieutenants, is at PT. If you can run, people make positive assumptions about your ability. If you can't run . . ." She shrugs.

Sometimes LeBoeuf has to keep women from pushing too far, as with the two women cadets who asked to join the boxing club. The male officers who oversee the club—and who are responsible for the preparation and safety of the cadets—came to LeBoeuf on behalf of the women.

"This captain is hardworking and sincere and wants to do a good job, but he's really not an expert in boxing," LeBoeuf says. "So I had to take that into consideration when I listened to his advice. He thought the women were ready."

"So I have to ask myself, 'What's next?' They'll probably want to travel with the team and box outside West Point. So then I have to run this past the 'front page of the *New York Times* test.' If one of these women gets hurt, and the newspapers ask if we did everything possible to prepare them and set them up for success, what kind of answer can we give them?"

"So I thought about what we do to prepare men—most of whom have absolutely no experience—for boxing. We give them nineteen lessons in a structured program. So that should be good enough for the women, too."

LeBoeuf contacted her boss, Brigadier General Abizaid, and asked his thoughts about having women box. Abizaid told her that he trusted that she would make a good study of the question and come up with a sound decision, which he would support.

LeBoeuf decided that any woman cadet who wanted to box would first have to audit the nineteen-lesson plebe boxing course. They would only spar with other women in that time. Successful completion of the plebe course is the only prerequisite for a man who wants to climb into the ring in intramural boxing; so it is for women, too.

The boxing ring isn't the only place seeing change. Two women wrestled in the Brigade Open Wrestling Championships, a contest open to the entire corps.

"These women had wrestled for four years in high school—against men—in the 123-pound weight class. They had some experience."

"The men in the crowd really behaved professionally," LeBoeuf says of the popular event. "In fact, I heard two cadets talking about the women's match later, and one of the guys said, 'It was a better bout than some I saw.' "

"I talked to those two women afterwards. I told them, 'You made history tonight.' "

Women's performance in athletics and physical education is not an academic exercise for LeBoeuf. Nor is it about winning a victory for feminism. It's about the Army living up to its promise to have a force that benefits from the diversity of its members.

LeBoeuf recalls interviewing cadets for chain-of-command positions.

"I was talking to one of the women about Close Quarters Combat," LeBoeuf says. "And I asked her for some feedback. She looked me right in the eye and said the course was good, but that women needed to learn how to attack."

The cadet was not advocating women in the infantry. She was talking about women taking the lead.

LeBoeuf is heavily involved in recruiting instructors for her department, in part because she is aware of DPE's reputation among cadets and graduates. She was not happy with what she saw—in terms of leadership—during her first assignment as a physical education instructor. Too much yelling, too many personal attacks.

"I tell my faculty to treat cadets the way they want cadets to treat soldiers."

LeBoeuf believes in the Commandant's gospel of hard, fair, respectful treatment of subordinates, who are held to a high standard. To keep her faculty form backsliding to the bad old ways, she holds team-building exercises in places where not everyone is comfortable.

"Last time, we went ice-skating. Now you take all these gifted athletes out on the ice—and lots of them can't skate. So they're lined up along the side of the rink with white-knuckled grips, like little kids in a skating class.

"Another time we all went swimming. You take someone like Major White, who's this big, muscular guy, a stud athlete in the boxing ring and on the football field, a guy who scares cadets just by looking at them. Put him in the pool and he needs two flotation vests to keep from going down, and he's one step ahead of sheer panic.

"So I tell them, 'You only teach what you're good at; cadets have to take a little bit of everything. I want you to remember how you felt, how scared you can still be. That's how cadets feel when they come into our classes.' "

LeBoeuf's assistant carries in neat folders of material to be read, papers to be signed. There are appointments scheduled throughout the day; she must prepare for tomorrow's trip to Washington and staffing meetings at the Pentagon. The guest speaker for a conference West Point is hosting for half a dozen universities just canceled. But for the moment, she focuses on what this assignment means to her.

"It's great; it's what I wanted. It's good for my family," she recites.

"But while cadets are great, they're not soldiers. My buddies talk about their troops and what it feels like to lead people in the field. I love my captains and majors, don't get me wrong, but it's not the same thing as having that [junior enlisted] driver who greets you in the morning with a big smile, even when it's a miserable day in the field."

She pauses. In the conference room next to her office, there's another meeting going on.

"I wonder if I'll get itchy feet after a few years here; I wonder if I'll be ready to move again."

THE HARDER RIGHT

*Encourage us in our endeavor to live above the common level
of life. Make us to choose the harder right instead of the easier
wrong, and never to be content with a half truth when the
whole can be won.*

from The Cadet Prayer

A cadet will not lie, cheat, steal, or tolerate those who do.

The Cadet Honor Code

In a speech to the yearling class, Retired General Norman
Schwarzkopf told the assembled cadets about his first week as the
number-two man overseeing the Army personnel system. One
morning he came into work to find his boss leaving, briefcase in hand.
Schwarzkopf, unsure of the new job, started firing questions at the
three-star. His boss turned to him calmly and said, "Follow rule number
fourteen."

"What's that?" Schwarzkopf asked.

"When placed in command, take charge."

This wasn't enough for Schwarzkopf. He followed his boss down the hall, badgering him with questions about upcoming meetings and decisions that had to be made and programs and news briefs and on and on.

"Rule number fifteen," his boss said.

"Rule number fifteen?"

"When in command, do the right thing."

The story's punch line was Zen-like in its simplicity, and about all the retired Schwarzkopf could do in fifteen minutes with a room full of nineteen- and twenty-year-olds.

Speeches by visitors such as Schwarzkopf are meant to reinforce messages the cadets hear constantly. Do the right thing. Live honorably. Duty, Honor, Country. But mantras aren't enough. There is a structure in place to educate cadets on the nuances of what these aphorisms mean, and one of these programs is built around the cadet honor code.

Honor education is the province of the cadet Honor Committee, which designs the four-year program with the help of the officers from the Commandant's office. There are formal classes, lectures, company meetings, and informal discussions. In an attempt to raise the standards of honor education, West Point established the Center for the Professional Military Ethic in 1998.

Lieutenant Colonel Charly Peddy, Deputy Director of the CPME, says that while honor education has long been a part of the cadet experience, it was uneven and sometimes haphazard.

"No one really sat down and thought about how we should develop character among the cadets. It was almost as if it was supposed to happen by osmosis: Put cadets among a bunch of staff and faculty and hope the stuff rubs off on them."

The center will develop lesson plans, integrate the education over four years, and generally seek to standardize cadet development. One of the initiatives in this area is the establishment of CHET, or Company Honor Education Teams. Two or three faculty members are assigned to work with each company, to help the cadets develop lesson plans for honor education. The idea is to give the cadets the benefit of the years of experience residing with the staff and faculty.

Colonel Peter Stromberg, head of the Department of English, is a member of the Company Honor Education Team for Company D-3. Stromberg, USMA '59, is the senior department head at West Point. He is tall and thin, with a deep voice, a keen sense of the absurd and an ability to laugh at himself. He is in Thayer Hall to observe an honor class conducted by the yearlings of D-3.

Inside the room, twenty-some yearlings mill about, trying to patch together short skits that will illustrate the learning points laid out for them in the prescribed curriculum. After a few minutes delay, the second class honor rep opens the door and invites Stromberg and two other faculty members of the CHET team inside.

In the first skit the yearlings create a hostage drama, in which a police chief must decide whether it's OK to lie to terrorists holding hostages. His plan is to tell them he wants to negotiate, then, when they let down their guard, send in the SWAT team. Another police officer questions whether this tactic is justifiable.

Although this group of performers spent time preparing (they had even typed out a script that sounded a lot like a Bruce Willis movie), they confuse the audience. Some cadets claim that lying to terrorists is OK, others aren't sure. Ultimately, there is no resolution.

The second skit involves three cadets in a firefight with enemy soldiers. Pinned down behind some desks, almost out of ammunition, the three consider surrendering. One suggests they hold up a white flag, then shoot the enemy soldiers when they reveal themselves. True to the lesson plan, if not very plausible, another cadet argues calmly that this would be a violation of the Law of Land Warfare. The cadets watching laugh at the performance. On this day, safe in the classroom, the skit has as much to do with them as space travel.

The cadets have stumbled through a half hour already. The CHET members sit quietly, and the audience keeps checking the clock. One cadet has his head down on the desk and might be asleep.

Stromberg whispers, "This is student-centered learning."

The last skit concerns something the cadets call the "home team, away team dilemma."

A male cadet, home on leave, sits at a table with a young woman. After ordering dinner, he suggests a bottle of wine that "might make things more fun later." His date giggles, demurs, then agrees. Later, we see the same male cadet back at West Point, where he runs into his cadet girlfriend. The two make plans for what passes as a big evening for West Point couples.

"Let's go running," the woman says.

"Great, and after that, we can go to the gym."

"Yeah," she says. "Lift weights, play some hoops. Then . . . uh . . . I guess we could go out for a run or something."

The conversation turns to what happened on leave; asked what he did for fun, the male cadet gives a half-hearted, "Not much."

A "good angel" appears by one shoulder, advising him to tell both women everything. A "bad angel" appears on the other shoulder, and with a few winks at the audience, advises him to keep his mouth shut.

"As long as you don't actually lie to either of them, it's OK," the bad angel says.

During the discussion that follows, the men in the room laugh and poke one another like high school boys. The talk centers around whether or not the cadet's actions, so far, have constituted an honor violation. No one asks whether he behaved honorably.

This is a common complaint among many cadets and faculty members: cadet concerns about honor aren't about honorable behavior, they are about whether or not some action is a violation of the code. The proliferation of rules, and one could even argue the nature of the litigious society the cadets come from, has created legions of guardhouse lawyers who are able to justify behavior that is outside the spirit—if not the letter—of the code.

One cadet, a woman, raises her hand.

"This wouldn't even come up if you respect both of these women. You wouldn't treat someone like that if you really try to respect people."

No one agrees or disagrees, and her comment passes.

"No, it's not an honor violation to have eight girlfriends," the junior leading the group adds. "In fact, it's pretty impressive. But it isn't right."

Recently, a cadet chosen by his class to chair the Honor Committee did not get the job because it came out that he was dating a cadet woman and was practically engaged to a woman back home. The two-timing cadet did not get the job, nor was he sanctioned by the Honor Committee, which has set a precedent of not getting involved in cadet dating lives.

"If we did get involved [in dating spats]," D-3's rep says, "there'd be a danger that some jilted girlfriend or boyfriend would bring someone up on honor for something we could never figure out: 'He said he loved me,' She said she wasn't sleeping with anyone else.' "

The cadets run their own honor classes and teach each other, exploring the questions raised at their own pace. Because they are inexperienced, because the presentations are ungraded (which means low on the priority list and not much prep time), the whole process is inefficient. As with leader development, it may be much more time-efficient for the senior leader to just say, "Do it this way," but then the lessons and the rules belong to the leader, not to the led.

But sitting in the classroom, with the clock ticking and every cadet aware that this is ungraded, that there are another half dozen graded requirements before the end of the day, makes it tough for the cadets to give this lesson its due.

Lieutenant Colonel Scott Snook, a leadership professor, says that real character development cannot take place in the classroom.

"We *train* people in the classroom. We can teach them the things a leader needs to know, and what a leader needs to do."

The Army's leadership model, laid out in the manual called *Army Leadership* is "Be, Know, Do." The "Know" part is skills and knowledge: it includes everything from operating equipment and preparing reports to planning attacks. The "Do" part is action. Leaders influence people, operate to accomplish the mission, and work to improve the organization.

But what a leader needs to be—the character issue—is much more complicated.

"We approach the character part like we approach everything else," Snook says. "Got a problem with lying? Throw a class at it. Sex discrimination? Hold a meeting. We approach it that way because we know how to train people."

To put it another way, when all you have is a hammer, everything looks like a nail.

Snook and others who want West Point to aspire to a more sophisticated approach say that cadets have to be ready to learn lessons about character. They must be self-aware and mature enough to be able step back from themselves and see how they fit in. Young people are most open to this kind of learning when something has happened to them that, as Snook says, "shakes their view of the world and how they fit in."

In my experience as a cadet, that meant failure.

By the end of my second year at West Point I was, by all objective standards, a successful cadet. My year of college experience before West Point prepared me well for academics. My grade point average and good standing in military attitude moved me high enough in the class to win the gold stars that indicated a top 5 percent class rank. I was focused and intense and probably not much fun to be around. For my summer assignment I chose the U.S. Army Ranger course, which was then available to cadets. Only a few cadets got to wear the black and gold tab; it was a significant achievement and would, I was certain, make a fine addition to my growing list of accomplishments. Cocky and underprepared, I went to Fort Benning and promptly flunked out.

I was stunned. I had never failed at any major endeavor in my life. At first I blamed the school's grading system, then the ignorance of the instructors, who obviously hadn't taken the time to be impressed by my lengthy credentials.

I suddenly had unexpected leave time on my hands. Somewhere in the course of those weeks before school started again, it occurred to me that the U.S. Army Ranger Department didn't owe me anything. I simply didn't meet the standard. That revelation made me think a great deal about how, in my two years at West Point, I had become a "sweat." I worried excessively about the smallest details because that's what West Point wanted, but I had no time or sympathy for people who could not perform to the same level.

In failing, I learned my own limitations and thus became more understanding of other people, critical lessons that I could never have learned in a classroom. Two years later I went back to Ranger School and not only completed the course, but I was able to help others get through.

While character development may be the trickiest aspect of the Academy's mission, the biggest challenge cadets face as they learn to live with the Honor Code is the non-toleration clause. A cadet who knows of an honor violation but does not report it is also guilty of a violation, and the sanctions can be just as severe.

"It's one thing to talk about honor in the corps in general, it's quite another to think about turning in a friend," says Colonel Maureen LeBoeuf, head of the Department of Physical Education and a CHET mentor. "[The non-toleration clause] seems to them to set up a tension between two goods: loyalty to their buddies and loyalty to the institution."

In another CHET class, LeBoeuf sits in the visitor's chair while a firstie named Belmont has the floor. He is trying to follow a lesson plan, and he isn't making much progress. After struggling for ten minutes, he asks if LeBoeuf will come up and share with the class a couple of things they spoke about in their one-on-one discussion. She agrees, and as she stands to walk to the front of the room, Belmont tries to thank her. But it comes out as, "I'm sure Colonel LeBoeuf will do a good job."

"Thanks Belmont," she says dryly, "It always makes my day to be validated by a firstie."

The class laughs, and she keeps the banter going, playing to what they will think is funny, joking with them, poking fun. She addresses one huge cadet; the desktop looks like a cafeteria tray in front of him.

"OK, now this is a stretch, are you a football player?"

"Yes, ma'am."

"And you guys learn to count on each other, right? You've got to trust each other, right?"

The whole room nods. They're all athletes, and they know the teamwork drill.

"It's the same way in a unit."

She walks over to two cadets sitting beside each other, weaving a story for them: The two men are friends. One, whose name-tag says Abelli, also knows that the other, Wade, borrows things from other cadets. Sometimes he "forgets" to return them.

"So he's not the most truthful guy," LeBoeuf says. "But then you find out you're going to the same unit, and this is great, because you're buddies.

"Then it comes time to report on your vehicles. And you, Abelli, give an accurate report. 'Eighty percent [mission capable] sir,' you say to your company commander. And the captain looks at you a little funny, 'cause that's not a great number. Keep in mind this is the guy who's going to write your efficiency report."

"And then he asks Wade, who saw the reaction you got, and Wade says . . ." LeBoeuf turns to Wade, pauses.

"Work with me, Wade."

"One hundred percent," Wade says.

"Right."

She turns to Abelli, then to the rest of the class. "And you know that's not true. What does that do to you? What are you going to do?"

Abelli doesn't answer right away. He looks a bit confused. Finally, he says, "I'd probably report 100 percent next time, too."

"*No!*" LeBoeuf insists. "No, no, no!"

Abelli looks pained. Not only were 20 percent of his vehicles broken, now he's given the wrong answer.

"He wouldn't trust Wade," another cadet helps out.

"*That's* right," LeBoeuf says. She looks unsure of what to do with Abelli; she sizes him up, wondering, perhaps, if he really would falsify a report that quickly.

"Suppose you and Lieutenant Belmont are in the same unit," she says to Abelli.

Belmont, the firstie, wears wings above his name-tag; he will become an Army aviator after graduation. LeBoeuf is also a pilot.

"And suppose Belmont tells you that this helicopter you're about to fly—which has been having problems—is now OK."

She leans forward, and in a stage whisper says, "But it isn't."

She straightens, addresses the class.

"How long do you think that helicopter will fly if the rotor bumps the mount?"

The cadets shake their heads in unison. This is working better.

"About as long as it takes to hit the ground," she finishes.

"We're not talking about working for IBM, folks," she says, moving inside the horseshoe of desks and looking each cadet in the eye. "You could go to war with this person on your left or your right. You've got to be able to trust this person."

These kind of extreme examples, combat stories and "what-ifs" are often heard in discussions of honor. But they are not the sole reason for the code.

When Sylvanus Thayer became West Point's Superintendent in the early nineteenth century, he instituted strict accountability for cadets: they were held to high standards in academics, in appearance, in comportment. Above all, they had to be gentlemen; that is, their word was their bond. Thayer knew this was essential in a society that mistrusted a professional military. There must be no opportunity for the Army's civilian overseers to say, "I told you they weren't to be trusted."

Cadet Kris Yagel, the senior honor rep in Rob Olson's company, says, "For me, learning the Honor Code started with fear; we were afraid of doing something wrong and getting thrown out."

Yagel, from Pennsylvania coal country, has dark hair and chiseled features. With longer hair, bad posture and a three-day growth of beard, he could be a Calvin Klein model. He is excruciatingly polite and serious about the Honor Code.

"It wasn't until cow [junior] year that I really began to respect the code and what it can do for me. This year I've been involved with the education committee, making plans for next year. Like, how do we get across that honor applies everywhere: to academics, to athletics, to cadet life?

Honor education for the plebes centers around how the code applies in cadet life, while yearling classes discuss honor in the military. The first two years are "directional", cadets are told what to do. By the end of the third class year and the beginning of second class year, the Honor Committee introduces more complex ethical dilemmas to the education.

"Not every situation has a right answer," Yagel says. "Some are cut-and-dried, like a falsified status report or an illegal order. Others are trickier. [But] the biggest threat to the Honor Code is the non-toleration clause. It's the most talked-about issue; it's the one cadets have the most problems with. I always tell them that, as an officer, you cannot afford to overlook dishonest behavior."

"Lots of cadets put loyalty to friends above loyalty to the organization. I used to think that way; that if I had a best friend who did something wrong that I wouldn't be able to confront him or turn him in. But now I see that the non-toleration clause is an important part of the code. Dr. Snider [Don Snider, a civilian professor and CHET member] talked about this once: Loyalty to the higher ideals, loyalty to the unit, is going to benefit everyone more in the long run."

There is another reason for the toleration clause: the code belongs to the cadets. If cadets don't enforce it—fairly, even in the smallest circles of friends—then there is no code. Still, Yagel has had other cadets tell him that they would not turn in a friend.

"It's tough. We're always taught: have a buddy, count on your team. You're always depending on someone. Now I have to turn my back on that person? It's hard to understand."

Visitors to West Point almost always comment on the openness: unlocked doors, unmonitored exams. But it isn't a perfect world.

"There are other problems, like 'cadet borrowing.' One of your sweatshirts is missing, which means someone took it, and then it doesn't ever come back. They say it's civilians in the barracks, but it goes on too much.

"So I ask people, 'What do you think was going through the mind of the cadet who took this sweatshirt?' People will say it's petty, but if you're going to lie about this what are you going to do when the stakes are high?'"

Yagel has a special disdain for cadets who won't take responsibility for their actions. He saw this when he served on the board that investigated incidents of electronic copying.

"I got so tired of sitting on those boards and hearing people say they didn't think [electronic copying] was cheating. It seemed clear to me. Maybe I'm getting cynical, or turning into an old grad, but it used to be that if you committed a violation, you got kicked out."

During Christman's time as Superintendent, there has been a sea change in the disposition of "found" honor cases. The code used to have a single sanction: Cadets who were found to have violated the code were dismissed. But Dan Christman believes in rehabilitation, in a learning curve, in a developmental model that allows for some missteps. He uses sanctions other than separation in some cases: cadets may be turned back to the next class. A junior or senior may be sent into the Army as an enlisted soldier; after some period (assuming good performance), he or she is allowed to re-apply for admission to some later class. Many of these cadets also participate in "honor mentoring," an intense program of one-on-one meetings with a senior faculty member. The cadet is required to keep a journal (for self-examination), to write papers on ethics, to read, even to address classmates on what he or she has learned.

A lot of old grads are unhappy with the change. They believe Christman has somehow sold out a Camelot they remember for an academy that allows cheaters to graduate. Such black-and-white memories gloss over other features of the single-sanction system: a board of cadets that wanted to take extenuating or mitigating circumstances into consideration might do so by voting "not guilty." And of course, West Point never was a perfect world: witness the cheating scandal of 1976, or a 1951 scandal involving widespread cheating over a period of time. (According to author Bill McWilliams, USMA 1955, eighty-some cadets eventually resigned in the wake of that scandal. The investigating board blamed it on football players; McWilliams says that was the simplistic finding of a "deeply flawed" investigation.)

"The new Supe has changed things," Yagel says. "It's hard to go from 'You did it, you're out,' to 'You did it, you're still here.' But I understand why.

"I was fortunate to have parents who taught me these values early on. But for many cadets, this is their first opportunity to think about these things. Why have such strict consequences if this is really supposed to be a place for learning? They get introduced to the concept, then they internalize it. If there isn't an environment where you can make a mistake, how are you supposed to learn?"

Yagel says most of his classmates support the single sanction: you did it, you're gone. The hard-liners are, amazingly, some of the same people who resist the non-toleration clause. The problem comes down to this: it's easy to be hard-hosed about honor in the abstract, when the cadet is a stranger, or from some other class. Yagel takes his responsibilities very seriously; he knows that the Honor Board is making decisions that will affect a young man or woman's life. Not all cadets strive for such balance.

"You have your 'Honor Nazis,' who want to hammer everyone. They go into these boards thinking if you're here you must have done something wrong."

Being an Honor rep, Yagel says, "is kind of awesome." His approach to the classes, his fairness, and even the example he sets influence the entire company's acceptance of the Honor Code. That code, Yagel believes, prepares cadets for life.

"It prepares you to make ethical choices. The goal is not a black-and-white view of right and wrong; the goal is to be an honorable person."

It is late on a winter morning, and Lieutenant General Dan Christman is also thinking about the Honor Code.

Outside Taylor Hall, winter sunlight bounces off snow and river ice but does nothing to warm the air. Christman crosses the foyer outside his office, walks through the Academic Board Room, with its ten-foot fireplace and conference table heavy enough to land aircraft on. The briefing room is huge: the barrel vault ceiling thirty feet high. State flags hang from the walls; arched windows let in bright winter light. The tables are oak, the floor is oak, the walls paneled in oak.

Christman greets everyone in the room, smiling and calling people by name. Brigadier General John Abizaid is here, along with a few members of his staff. There are military and civilian faculty members, and senior cadets from the Honor Committee. They all stand at attention until he directs them to sit. Christman has asked for an update on several studies looking at the cadet honor system and the state of honor and ethics training in the Corps of Cadets.

The cadet Honor Code belongs to the cadets; throughout the briefing Christman gives the floor back to the cadets again and again, gently taking the discussion away from the nine officers and one civilian professor sitting at the table. Besides preserving cadet dominion over the system, this gives him the best sense of attitudes within the corps.

During the discussion, someone puts up a slide that shows the definition of electronic copying. This has become the latest hot topic for the Honor Committee. Someone suggests that the definition should be further clarified. Colonel Kerry Pierce, the Superintendent's chief strategic planner, claims that a clearer definition won't be enough.

"The vaguer their understanding, the safer they feel," he says. "Just because they've been educated doesn't mean they have a desire to understand the definition fully. A lot of the cadets who wind up accused of electronic copying see claiming to be confused as a tactic. Cadets try to stick with the letter of the law when it suits them."

A civilian professor on the committee says that some cadets see some honor violations as "petty," on a par with violating some small regulation.

One professor says the endless pages of regulations keep the cadets from assuming responsibility. They think that getting away with a breach of regulations is part of the "game", that it shows some spirit. Some cadets transfer this same mentality to the cadet Honor Code.

"They don't want to know what the code says so they can believe they didn't violate it," another professor says.

Christman seems amazed, even a bit exasperated.

The overwhelming majority of the Corps of Cadets lives by the Honor Code; most of them see it not as an onerous imposition, but as a perquisite that comes with cadet life.

One cadet sums it up as, "It's nice to be able to trust people."

But the cadets who follow the rules aren't the ones Christman sees. As the final arbiter of honor cases, he is presented with a sad procession of bad decisions: a cadet who used a fake ID card to get into a bar at Fort Benning; another accused of shoplifting. One or two cadets who copy work, then fail to document the assistance. They misrepresent themselves, they quibble, they shade the truth.

"Why would a cadet risk a failure of integrity just to avoid an academic failure or a bad grade?" Christman asks the big room. "We probably have hundreds of cadets who have failed a couple of academic courses and have made them up. There are no cadets around who have broken the Honor Code twice."

Christman does not claim it is an easy code to live with; but it is the only way to run the army.

One professor says that the explosive growth of information available on the Internet—such as term papers that can be downloaded for a fee—tempts some cadets. But Christman interrupts. He isn't worried about temptations facing the cadets because he can't control those. He is not in pursuit of a perfect, insulated world. He is in pursuit of graduates who can exist in the murkier waters of the world beyond the gates and still have the courage to make the right choices. The answer, as he sees it, lies in education.

"We've got to do everything possible to make sure cadets recognize ethical decisions. We can't mandate decisions, no matter how many rule books the Honor Committee or this committee draws up."

"Part of this mind-set [of always checking the rules] comes from the fact that accused cadets consult with lawyers," Christman says.

"And a good lawyer is going to look for loopholes, is going to try to get his or her client out from under the accusation."

While the lawyer Christman knows this is good for due process, it works at odds with one of the great goals of the entire honor system. Looking for loopholes, technicalities, process errors, all those things encourage the cadet to avoid taking responsibility. And that is a cardinal sin in character development.

"Some of these cadets are found, they admit that they did it, yet they still refuse to accept responsibility," Christman says incredulously. "Anything except take responsibility."

In a crude way, this is what the Academy tries to teach by giving new cadets only four responses: "Yes, sir," "No, sir," "No excuse, sir," and "Sir, I do not understand."

Give that third response, and a cadet can expect to be punished. But when that eighteen-year-old plebe learns to use it for what it really is—an admission of personal responsibility—he or she also gains a measure of respect that is indispensable for a leader.

Christman cannot legislate this; he cannot order it. He can train and educate; he can inspire. He must. Acceptance of personal responsibility is a hallmark of the officers corps, part of what defines the profession. Soldiers reserve a special disdain for those who won't accept responsibility.

Christman, against all conventions, against the norms for society, against everything played out on the news, against the example set by the commander in chief—is trying to teach something else.

Another subcommittee puts up a slide showing the results of a survey conducted among members of Corps Squad teams. The study tried to identify whether the cultures of certain teams affected the way cadets on those teams viewed honor and ethics.

There are two parts to the study. The first surveyed cadets' ability to recognize an ethical quandary; the second showed their desire to live honorably. In both studies, women's teams scored as high or higher than the rest of the corps. In both surveys, the hockey, football, power-lifting, and lacrosse teams scored below the norm.

Several theories are offered around the table: football and hockey are "heavily recruited," one officer says. The Academy goes out after kids who play these sports, it could be that players on these teams are not as self-selecting as the population of cadets in general.

"We have to consider that coaches come from different environments," Christman adds. "Where winning is the most important thing. We've got to get to them early, make sure they understand that ethics is the heart of what this institution is about. Winning is important; this is more important."

"One or two cadets can adversely affect a whole team," the Superintendent adds. "We have to change this through the chain of command."

Christman has directed that all team captains also be made cadet captains; they wear the same four stripes as cadets who command companies. They are invested with some authority and, as in any military organization, they are held responsible for their people. This is not just a Band-Aid. Team captains—all of them seniors—now must answer for the conduct of their players on and off the field. Christman has superimposed a military structure on top of any informal structure that might have existed, where the best players, the starters, might be the only team leaders.

Another slide goes up on the screen, a vast array of numbers. A lieutenant colonel takes hold of the electronic pointer.

"We also looked at the treatment of minorities in the honor system," the briefer says.

With the help of the math department, which did a statistical analysis, the committee found that treatment of minorities does not differ in a way that is "statistically significant."

Abizaid points out that his staff has done a similar study of punishments for breaches of regulations and found that African Americans and women are not treated differently.

Christman warns the people gathered around the table that they need to keep these facts and statistics at hand. A recent newspaper article about a suit filed against the U.S. Naval Academy makes Christman think that West Point will soon be getting inquiries. The *Washington Post* article detailed an honor investigation at Navy; it also showed just what kind of Gordian knot can land on the desk of the Superintendent on a Monday morning.

According to Major John Cornelio, USMA Public Information Officer, a male and a female midshipman were accused of having sex in Bancroft Hall, the midshipman dormitory. The punishment is often expulsion. The male freely admitted to having consensual sex and was forced to leave the Academy; the woman claimed it was date rape. Other midshipmen witnesses said there was no sign of compulsion, that the woman was inebriated but not resisting. The investigating officer appointed by Navy's Superintendent found that it came down to a matter of he-said, she-said. There was not enough evidence to accuse the male of date rape. On the other hand, Academy officials could not dismiss the woman (for having sex in the midshipman dormitory) because they could not prove that the sex act was *not* consensual. Nor could Navy retain the male midshipman, who admitted to breaking a rule that calls for dismissal. The sex act thus existed simultaneously as consensual sex *and* date rape.

The woman was allowed to stay at Navy. The male midshipman was dismissed. Because the woman is white and the man black, his family maintains that the decision was evidence of racial prejudice.

Throughout the meeting Christman compliments the work of his subordinates, especially the cadets, who have done several studies. Cadets on the Honor Committee put in long hours: watching the progress of investigations, seeing to the education program, sitting on boards, conducting investigations, and talking to cadets. All of this is on top of their regular course work.

Christman leaves the cadets with an admonition not to let down in the few months before graduation. It is their responsibility to ensure that the work of the Honor Committee continues smoothly as the next class takes over. His responsibility—nothing less than the moral development of a generation of Army leaders—has a longer horizon. West Point takes this deliberate, detailed approach to moral development because of the explicit statement, in its mission, to create "leaders of character."

WE'VE NOT MUCH LONGER

We've not much longer here to stay
For in a month or two
We'll bid farewell to "Kaydet Gray"
And don the "Army Blue"

Army Blue
By tradition, the last song played at West Point dances

Kris Yagel, the Honor Rep for Company E-2, sits in the First Class Club, nursing a beer and shaking his head at how quickly his senior year is drawing to a close. "I've already turned over my Honor Rep duties to the second class in my company. Ten days of class, then exams, and then . . . it's graduation," he says.

Cadets spend much of their time counting days and wishing huge chunks of time could just disappear from the calendar, but now Yagel is feeling a pull in the other direction.

"I can't wait for graduation to get here, but I also want to spend time with my friends, because we're about to be separated."

At tables all around him, Yagel's classmates seem determined to get the most out of this Friday night.

The Firstie Club is part college bar, part shrine. Built as the Ordnance Compound in 1840, the building was used to store ammunition and powder (ordnance) until the twentieth century. The walls are covered with photographs documenting the cadet days of the Class of 1958, which renovated the club so that the seniors would have a place to relax. In the not-so-distant past, cadets were forbidden to drink within twenty miles of West Point, which of course meant they just drove to bars farther away. Now twenty-one year-old seniors can drink beer within walking distance of the barracks. Many old grads think that such a privilege is apostasy, but Academy officials found that allowing cadets to exercise some of the same privileges enjoyed by their contemporaries at civilian colleges makes them better prepared to handle the sudden freedom that comes with graduation.

There are dozens of tables and a smaller room filled with pool tables, video games, and a jukebox. In the forty-five-year-old black-and-white photos, the athletes wear bulky, high-topped athletic shoes; the buildings visible in the background belong to an older West Point. The young men in the photos, frozen in their season, look like Kris Yagel: scrubbed, healthy, earnest. None of them would have believed the years could speed by so quickly; Yagel is feeling a bit of that amazement.

Across the table, Kevin Bradley and a couple of other firsties sit talking with Chuck Ziegler, a retired Army officer here for a weekend visit. They ask Ziegler, a former infantryman, about the force he joined twenty years earlier. The Army he describes—Ziegler served in Panama, Honduras, the Dominican Republic, Korea, Hawaii—sounds as exotic to these cadets as the wars against the Plains Indians.

Bradley and a few others drink Coke. They are only hours away from the start of the Sandhurst Competition, a grueling marathon of military skills that has dominated their lives for the past three months. The next morning will be filled with obstacle courses, running, land navigation tests, running, weapons disassembly, and assembly, running, shooting, running, rappelling, and more running. Bradley's parents are even coming to watch. But the talk around the table is of the Army, of what awaits them, how well they are prepared, and what decisions they must make. Bradley has reached at least one decision that weighed on his mind during the winter. He isn't getting married.

"Just not ready, yet. I'm twenty-one years old and haven't lived on my own."

He and his girlfriend of a year, Amy, understand that his move to Fort Knox, then to Europe, will mean the end of the relationship.

"She's getting ready to go to Australia for a semester in the fall," Bradley says a little glumly. He is already beginning to feel the cost of the turmoil that comes with military service: frequent moves can be tough on a personal life.

Behind Bradley, another firstie, a woman in the spring uniform of dress gray coat and starched white pants, collects empty plastic cups and trash from the tables. She has the duty for the evening. Ziegler points out that the bar is neater than one would expect to find at most student unions.

"Course, the guard might have something to do with that," he jokes.

Kris Yagel's first assignment is Fort Polk, Louisiana, where he will be a Military Police lieutenant. MPs are either in law enforcement (the on-post police force), or in tactical units. Yagel is going to a tactical unit, which means frequent deployments overseas for peacekeeping missions. He and his soldiers will patrol civilian areas, maintain traffic and transportation routes, and generally enforce the *Pax Americana* at the beginning of the new century.

The discussion of overseas deployments brings the talk at the table to Kosovo and the prospects for a ground war. The possibility is on everyone's mind here, especially the firsties. No graduating cadet will go immediately to the Balkans; all of them will first go for five or six months of training in the technical skills they'll need to command. But many of them will get there soon enough.

Bradley is slated to join the First Infantry Division, the famed "Big Red One," in Germany. A month before the conversation in the firstie club, three soldiers from the division were seized by Serbian forces from a road in Macedonia. This evening, as Bradley and Yagel sit in the Firstie Club, the Serbian government is threatening to put the three on trial.

That spring, when the air campaign against Serbia began, Bradley and the other firsties followed the news closely. Cadets are not privy to national strategy, of course, and they do not have any special training to help them predict what might happen. But they are educated people who read newspapers, and they're interested in current events that will affect them personally.

Bradley believes he will go to Bosnia at some point. In fact, that was one of the reasons he chose the First Division, whose soldiers make up a good portion of the force in the Balkans. (Bradley, with his high class rank, had his pick of any assignment offered to the Class of '99.)

"I think it will be a good place to learn my profession," Bradley said. "A good place to learn how to be a platoon leader. I mean, I'll actually be doing the job I'm supposed to do."

His choice is ironic, given that he had been involved in a student debate on U.S. policy on Bosnia while he was in high school. He was opposed to American intervention and argued strongly against it, but now he sees government policy as something that generated missions for the military.

The first class, not surprisingly, spends a lot of time talking about what is happening in the Balkans. Bradley tells the group that his roommate, an Army soccer player, "never really spent a lot of time thinking about what it will mean to be in the Army. Now he's going to the 82nd Airborne, and he thinks about it a lot."

A recent newsmagazine photo showed soldiers from the 82nd Airborne Division in Albania, where they were to provide security for the Apache helicopters that joined the NATO force in April.

"He's got it taped up in the room," Bradley says.

For all the talk of a messy involvement in the Balkans, the club is not a gloomy place; instead, it's filled with the buzz of approaching adventure. Cadets play pool and sing along with the jukebox. They wear baseball caps and jeans and T-shirts from other colleges and talk about what they're going to do on the sixty days of leave between graduation and the beginning of training. They are on the verge of great changes, new adventures, and more hard work, and the very air around them crackles with anticipation.

The next day, Saturday, is like a postcard of spring. By 7:00 A.M. the sprawling barracks are humming with activity. Cadets in BDUs and others in running gear pass. One goes by on roller blades, looks up from the rhythmic movement of his feet on the pavement to greet an officer. "Mornin', sir."

The mountains on the east side of the river are lightly painted in greens and yellows. Through the open barracks windows come the sounds of cadets getting ready for Saturday morning inspection of rooms, the most thorough of the week.

Pete Haglin, the plebe determined to be an artillery officer, gets off a yellow school bus in front of Lee Barracks. He and two dozen other sleepy cadets in BDUs have just come from the start line of the Sandhurst Competition, where they cheered their company team. Haglin is happy that the semester is drawing to a close, but his earlier worries about whether or not he could handle West Point academics were right on target. His summer leave hangs in the balance; he may wind up in summer school for chemistry.

"I've got about a D minus minus right now. But I figure if I get about 80 percent of the points left [on the final exam], I'll pass with a C."

Summer school, called Summer Term Academic Program, or STAP, can eat up a cadet's entire leave. If he goes to STAP, Haglin will begin classes before graduation day and will stay at West Point through June. He'll get a long weekend off—assuming he passes his second go at chemistry—before reporting to Camp Buckner for summer training with his class.

As late as the 1970s STAP veterans were "awarded" a black star to wear on their cadet bathrobes. It was an ironic badge of honor, the equivalent of an academic Purple Heart (the medal given those wounded in combat). Eventually, the academy leadership decided that such an award did little to foster a serious approach to study. Black stars are no longer handed out.

Two firsties from Kevin Bradley's company, F-2, are among those coming out to cheer on their team. Nick Albrecht is from Ohio; he wears the crossed cannons of the field artillery, the branch he will join in a few weeks. Ron Havener, who is from the far north of California, wears crossed rifles; he'll head to Fort Benning, Georgia, the home of the infantry, for his training. Albrecht mentions Captain Brian Turner, their Tac.

"He had us over for dinner the other night," Albrecht says. The seniors laugh as they describe Turner's apartment, which is decorated in bachelor simplicity.

"He's got a couch, a recliner, and a stack of green plastic chairs," Havener describes. "We practically had to bring our own silverware."

If the two aren't impressed by Turner's decorating skills, they are impressed by his hospitality.

"That's the first time I've ever been to a Tac's house for dinner," Albrecht says. "In fact, Captain Turner is the first Tac I ever had who sat me down and really talked to me about the Army, about what to expect, about the connections between what we're doing now and what we'll find out there."

They drive along the highway that splits the big reservation. This is the same road that points to Lake Frederick, where classes of new cadets have ended Beast Barracks for fifty years. The road is lined on either side with signs, shaped like shields, bearing the names of the various ranges and training areas. They are named for American battles: Buena Vista from the Mexican War, Normandy Range. After a few false turns, they catch up with F-2's team at the rifle range.

The nine cadets on the team—Bradley is one of them—are camouflaged and outfitted in a field uniform: BDUs, helmet, rifle, protective mask, climbing rope, gloves, canteens, ammunition, and first-aid pouches. The squad carries one radio, with the cadets taking turns carrying it. One rucksack contains a climbing rope. The F-2 team passes their initial inspection; they will be inspected again at the end to ensure that they finish with the same prescribed load of equipment.

The team is fresh, well conditioned, and nervous. This crew includes four firsties, which may explain why they make decisions as a group, like something out of a business school study of how teams should work. The firsties have all been in this competition before. The cadets of the lower classes defer to those with more experience (as opposed to those with more seniority); but no one, not even the lone plebe on the team, is excluded from the decision-making process.

(Interestingly, this style of team leadership is used by the service's most elite units. Members of the Army's counter-terrorist unit, the Delta Force, are task-organized for missions; the most experienced, best-qualified man is in charge, regardless of rank.)

After inspection, the team draws ammunition for the first test: rifle marksmanship. When the clock starts, they run a quarter mile to the firing range (they won't stop running until they reach the finish line several hours later). More than sixty members of the company have turned out on this Saturday morning to cheer on the team. This is the majority of those who are not competing with or supporting Army athletic teams in away games.

The cadets who turn out to support the team wear a variety of outfits. This is a spirit mission, and the rules are always a little flexible when it comes to spirit missions. There are F-2 T-shirts with a cartoon gladiator, F-2 intramural uniforms, with the company designation. Other cadets wear combinations of camouflage trousers and "spirit" T-shirts; still others are in PT gear.

On the range, the rifles begin to crack. The targets, shaped like the head-and-shoulders silhouette of a man, pop up and get knocked down. After a few well-organized minutes they come off the range, have their weapons checked, and take off running.

The crowd follows the team along the trails that top the ridge-lines. The lead cadet in the cheering section carries a black four-by-six-foot flag with "F-2" in big gold letters. There are a half dozen cadets on mountain bikes, a few taking photographs. It is a beautiful spring morning, all sunlight and fresh air, team identity, and athleticism. The men and women would rather be back in bed, but sleeping in is not an option, so they cheer their company-mates.

Nick Albrecht credits Ron Havener with the team identity on display; Havener was F-2's cadet company commander first semester.

"Ron did a lot to pull the company together. He listened to people. He paid attention to the under classes. He tried to make things fun."

The squad moves at a fast trot—helmets, rucksacks, and weapons jangling—up a stony, eroded trail to a clearing where they stop and put on the thick protective masks. The black-rubber-and-plastic headgear makes breathing and seeing difficult. The cadets will run nearly a mile in the claustrophobic masks, sucking air through filters designed to keep out chemical agents. By the time they reach the next site, they are seriously winded and sweating heavily.

The spectators, running without masks, reach the next site first. They stand behind a white tape barrier as their team lines up behind machine guns set on tarpaulins. This station is timed disassembly and assembly of the weapons. Calm reigns, and when one of the contestants is penalized for a tiny infraction, the others merely lay on the encouragement.

More running. The company commander, Murphy Caine, has been carrying the big flag up and down the mountain trails. He has painted a black "F" on one cheek and a "2" on the other. In his enthusiasm this morning, he painted the letter and number while looking in the mirror, so that when he went outside the company had a good laugh: the letter and number were backwards.

Several miles into the competition, the cadets approach a stream and go into a practiced drill. They drop the rucksack containing the climbing ropes and begin constructing a one-rope bridge. One cadet wades through the stream and knots the nylon line to a pole on the far bank. The others thread their end through a series of knots, creating a come-along; they pull it tight, heaving together like sailors. The cadets have all fashioned "Swiss seats," the tight girdle of line to which they attach a mountain climber's 'D'-ring.

A second class cadet does a pull-up on the line, which is some five or six feet up, hooks the ring, and spins until he hangs parallel to the ground. With quick hand movements he pulls himself across to the other side, hooks a leg over the line, and unsnaps the ring. The others follow; the metal snap links make a buzzing sound as they fly across the braided nylon rope.

The bridge site is where the families of some cadets have gathered to watch. Kevin Bradley's parents, Dave and Madge, and his younger sister Cori, have driven the two and a half hours from their home in southern New Jersey to be here. They chat amiably with the cadets, many of whom they know from four years' worth of trips to West Point.

Bradley grew up playing football and baseball at his suburban high school; his parents cheered for him in countless games. If it strikes them as odd that they are now cheering their son and his teammates in a competition of soldiers skills that includes shooting, it doesn't dampen their enthusiasm.

Cadets in the crowd along the stream bank chant "Zoo, zoo!" (for the "F-2 Zoo") as the team races over the shallow water. Once across, the contestants pull apart the bridge, stuff the rope into a rucksack, and head off in a file along a narrow trail. Thick briars tear at clothing, tug equipment, scratch exposed skin. They push through the undergrowth to a flooded storm drain that runs beneath the nearby state highway: four lanes and a median. A cadet says that the Commandant, Brigadier General John Abizaid, crawled through with the team just ahead of F-2.

The Zoo team plunges into the muddy water and, one by one, squeezes into the concrete pipe. A half dozen cadets from the cheering section, led by the company commander, follow, hollering gleefully as they emerge on the far side.

There is another long run, this one in wet boots and uniforms. It is difficult now to tell the cadets apart, even the men from the woman. They are all soaked, muddy, sweating; the camouflage paint runs in green rivulets down their faces.

Up another ridge on a paved road beside a small nineteenth-century cemetery, then along a dirt track that leads uphill (again) for another half mile. The cheering section is in full stride now, the numbers seem to be growing. It's a smalltown parade, a gaggle of fired-up college kids in running shoes and on mountain bikes. They exchange gossip and talk about possible scores and who went through the culvert and how many points the company has.

Near the top of the ski slope the team enters a clearing. The crowd following, all good soldiers, falls behind the white tape marking the spectator area. In the middle of the clearing is a fourteen-foot-high wall made of wide planks painted black. The front of the wall is decorated with a gold Ranger tab that stretches seven or eight feet across. There is no platform on top, just another plank set on its side on the telephone-pole frame. There is a sawdust pit at the bottom on both sides. The task is simple enough: The team must get over the wall with all their equipment and weapons.

The Zoo team has practiced this particular event several times a week for the past two months; and it looks it. Two cadets back up to the wall, squat down, boost the leaders to the top. Even for the tallest cadets, it is still a stretch. The cadets scaling the wall must pull themselves from a dead-arm hang to the top of the narrow planking, then get a leg up, then go over without simply falling. When a few are on the far side, two men pause at the top to hand over the team's weapons, the rucksack with the heavy radio, the bridge equipment. The crowd shouts encouragement and advice.

Kevin Bradley is the next-to-last man to scale the obstacle. His teammate crouches, back against the wall, and Bradley steps in his hands. The yearling pulls with both arms and launches Bradley upward; two cadets straddling the narrow top of the wall grab his hands, pull him up. But Bradley doesn't go over. He sits on the wall, his back to the last man, hooks his knees on the edge, then leans over backwards, dangling as if from a trapeze. His helmet and equipment all head for the ground; the straps and hooks and belts hit him in the face. One of the cadets straddling the top of the wall pins Bradley's legs to the top plank. Bradley is upside down, his head a good seven or eight feet off the ground, arms extended. The last cadet on the near side backs up and runs full speed at the wall, leaping at the last second, clawing, grabbing Bradley's belt, his harness. Using Bradley as a human ladder, he scrambles up, stepping on Bradley's armpits. His flailing boots smack Bradley in the face. Then he is up and over, as quickly as if he had climbed stairs. Bradley and the last man leap to the ground, and the whole team is across in just over a minute.

The crowd goes wild. It is the fastest time the Zoo has delivered all spring, but there is no stopping, and the team takes off at a run, the crowd still cheering.

They run on a ridgeline trail to the mountaineering site, where the Zoo team rappels down a steep rock face above the tennis center. The following crowd clambers down the hill, using a climbing rope as a handhold. They race to get ahead and be in place for the finish.

Breaking out of the brush along the road by the Cadet Chapel, members of the team check to make sure they have everyone. Sweaty, blowing hard, they scamper down the steps by the gym and run out onto Jefferson Road, which passes the Supe's house. As they turn the corner, they come up behind two wide-bodied humvees. In previous years, the final task was to push a jeep around the block on which Quarters 100 sits; the humvee weighs more than twice what the old jeep weighed.

The team starts off fast, pushing the big green vehicle toward MacArthur's statue. Around the corner, sandbags are laid end-to-end across the road. The eight cadets (one is inside steering) strain to get the vehicle over the bags. In front of Arvin Gym, they maneuver the humvee around some traffic barriers. By the time they come around the block, they are near the end of their strength. Stumbling, mute, they grab their equipment, do one last check to ensure that they are all together and have everything (leaving a piece of equipment behind incurs a penalty; leaving a teammate behind means disqualification). Then, incredibly, they sprint across the finish line.

The street is lined with spectators; the Zoo cheering section is hoarse, but manages a last roar. Once the team crosses the line, another judge checks their equipment for completeness, and they are finished.

It is 10:30 on Saturday morning.

The Zoo team goes off to shower, to tend their wounds, to scrub the camouflage paint off their faces and arms. The Bradley family heads to the grassy field where the rugby team practices. They've brought along a picnic for the entire Sandhurst team.

The Bradleys are in their forties, both schoolteachers, with an easygoing demeanor that makes it easy to picture them in a class-room. They are the kind of adults teenagers would talk to. Madge Bradley, whom Kevin resembles most strongly, has dark, almost black hair. Warm and friendly, she smiles a lot and speaks with a strong New Jersey accent, like someone from a Bruce Springsteen song.

"When he first came here I was worried that we'd lose Kevin," she admits. "But in fact what happened was that my family didn't get smaller, it got bigger."

Madge comes from a large family, and so is used to having a crowd around. Because the Bradleys live so close to Philadelphia and the site of the Army-Navy football game, their home has been a gath-ering place before and after the game for Kevin and his friends.

"She loves it," Dave Bradley says, smiling behind his sunglasses.

"We have kids sleeping on the floor, on the couches, in all the rooms," Madge adds. "My brothers come along and I tell them to bring lots of food . . . those kids can eat."

The Bradleys' van is parked at the edge of Clinton Field, close to the center of the cadet area. A hundred yards away, tourists snake through the rows of cannons on Trophy Point and pose for pictures with the Hudson as a backdrop. Dave Bradley unfolds a large table. Cori covers it with a tablecloth and the family goes into a little drill, as practiced as the one-rope bridge exercise. Two big coolers, one full of soft drinks and the other full of long sandwiches ("hoagies," in New Jer-sey); brownies and chips and napkins and pickles. They've been doing this for four years, and now they can hardly believe it's about to end.

Madge checks to make sure Kevin isn't around, then tells a story. When Kevin was six years old, the family visited Washington and went to the top of the Washington Monument. The elevator operator looked down at Kevin and asked him if he wanted to operate the lift. Kevin, wide-eyed and excited, said he did, and the man told him which button to push.

"Kevin pushed the button and that man announced to the whole elevator, 'Look at this man, only six years old and already serving his country.'"

Dave Bradley says that Kevin grew up wanting to be a pilot. At a college fair, Kevin approached the table for the Air Force Academy.

"The guy noticed Kevin's glasses and asked him about his vision. 'You have to have 20/20 vision to fly for us,' the Air Force guy said. And there was this guy from West Point sitting at the table right next to that, and he jumped up and said, 'You can fly for us!'"

The Bradleys let their son find his own way into West Point. Kevin visited twice; after one of the trips he told his teacher-parents that he liked the way the classrooms were run. The sense of orderliness appealed to him, but the parents disagree as to whether Kevin ever had a particularly hard time.

"He didn't like getting yelled at," Madge says.

"He wasn't much of a yeller himself," his father agrees. "Kevin was captain of a couple of teams in high school, football and baseball. And he was always a quiet kind of leader. One of the coaches told me that Kevin would listen to the other kids, then he'd talk. And people would listen to him."

"Cori told me I should have yelled at him more to get him ready," Madge jokes.

He did, they agree, have a rough start to Beast in 1995: Kevin wound up in the hospital on R-Day with a stomach virus. He called home in a panic that he was already falling behind.

"I told him to just take it one day at a time," Madge says. Then, after a pause, she adds, "I follow that advice in my life, too. When Kevin was a plebe I spent a lot of time worrying about stuff that never came to pass. So now I try not to worry. It's a waste of energy. I'll still worry some—I'm a mother—but I've learned that most of what we fear doesn't happen."

Kevin and some of the other cadets from the company team show up, wearing loose-fitting PT shorts and shirts and walking gingerly. Kevin has a long scratch on the inside of one thigh where brambles tore through his trousers. He reports that he has a welt, almost perfectly square, where his rifle gouged him between the shoulder blades as he was getting off the one-rope bridge.

The team digs into the meal laid out on the folding table, piling their plates high with sandwiches and brownies and chips. The underclass cadets are quiet; talk among the firsties soon turns to graduation week.

Kevin says that some parents are hosting a picnic on graduation day. The firsties have been invited to be "pinned" there. The new lieutenants will take the oath of office at Michie Stadium, then receive their diplomas. Later in the day, dressed in army greens, they will have their new gold bars pinned on. They can choose the time, the place, and even who will pin them.

Kevin winces as he sits on a cooler and flexes his legs, then lets himself smile as he talks about the ceremony. His parents, who are about to lose him again—first to training at Fort Knox, Kentucky, and then to the U.S. Army in Europe—busy themselves with the details. How can we help? Can we chip in for food and drink? What time?

Kevin is a little short on details because there are just too many other things to worry about between now and graduation day. He finishes his sandwich, puts his paper plate on the ground, and stretches his tired legs in front of him. Then he checks his watch.

"Gotta go," he says, pushing himself into an upright position.

He is, incredibly, off to practice diving with the Scuba Club.

"Thanks for lunch and everything, and for coming up," he says, kissing his mother's cheek.

The cadets say their thank you's and hobble away to the next requirement. In a few minutes, the Bradleys are left alone. Dave watches his son as Kevin walks away across the grass.

"He takes advantage of every opportunity to learn that they offer him," Dave says. Then he folds the picnic table and packs up what little food is left.

If West Point excels at anything, it is at taking advantage of learning opportunities. Early on a spring morning, Pete Haglin is experiencing one of these "developmental opportunities."

Haglin is the "section marcher" for his boxing class, responsible for getting the class ready: everyone in uniform, attendance taken, mouthpieces in, headgear on, lined up in two ranks before the instructor. Every general who went to West Point started out in charge of such a group. Haglin moves quickly to one of a half dozen lockers in the big boxing room and pulls out the equipment for class. Then he consults a pocket-sized notebook, calls names, and checks that everyone is properly outfitted. It is 7:30 in the morning, in the second semester of his freshman year in college.

The cadets pop to attention when the instructors enter the boxing room, which looks like a movie set for a period film. Built in the thirties, it is seventy feet long by thirty wide, with arched windows high in the front wall, exposed steel beams, and lots of worn brick. The lower parts of the walls are covered with thick yellow pads. A dozen heavy bags hang from chains at the front of the room, near the door. Tall mirrors stretch for twenty feet along the wall opposite the windows. The cadets move in a round robin of warm-ups, from the exercises to the heavy bags to the mirrors for shadowboxing. Then they gather at the end of the room where the ring waits.

Major White, the lead instructor, climbs into the ring. He is a big man, a former Army football player, handsome and muscled like a bodybuilder. White wears polished black coach's shoes, prim white socks pulled up above his ankles. His black shorts are, in the style of DPE, tight, like cut-off spandex. His shirtsleeves grip his biceps. The baggy look hasn't caught on here. His clothes look tailored, or deliberately shrunk.

The cadets have been moving continuously since the beginning of the class, going on fifteen minutes by the time they gather at the ring.

"I want you to get a little taste of what it's like to be a boxer,"

White thunders. "You're going to have a couple of puny little one-minute rounds in a fifty-minute class. Think about these pros, training five, six, seven hours a day, going round after three-minute round. Boxers are some of the best-conditioned athletes in sports."

Boxing is a required sport for all plebe men. The whole course lasts only nineteen lessons, but it has an effect all out of proportion to its duration. Plebe boxing is the source of a wealth of stories these men will tell for years to come because, for most of them, it is the first time they have had to confront real physical fear. The boxing ring, with its stained mat and unforgiving ropes, is where these future warriors learn about courage.

Haglin climbs into the ring. He wears a head guard with a bridge because he's broken his nose three times as a kid: once paying soccer, once playing basketball, once in a fistfight. But the bridge is in his way, and to see to the front he has to cock his head like a bird. He takes his corner, bangs his fists together as he shuffles from one foot to the other.

Haglin, who stands about five ten, has a good reach on his opponent, who is an inch or two shorter. But the other boxer is much thicker, maybe twenty pounds heavier. And apparently fearless. When White calls out, "Box," Haglin's opponent strides across the ring and delivers a series of jabs like jackhammer blows. None of them connects hard enough to stagger Haglin, but he gives up the initiative. His feet move too fast, his punches are a bit off the mark. Still, unlike some of the other boxers this morning, he is more concerned with landing solid punches than with simply avoiding getting hit.

"Good-looking jab," White calls out as the shorter cadet snaps Haglin's head back. "Use a combination now."

White expects them to think on their feet, to make quick decisions. The cadets know, intellectually, that they aren't going to be hurt badly (although boxing does cause a few concussions every round). Acting on that is another matter.

At the beginning of the second round, Haglin enters the ring with more determination. His opponent wades in gamely, though without finesse, and fires a couple of powerful jabs. Tired now, he also drops his arms obligingly every time he punches. Haglin sees it, takes advantage of the mistake and lands a few jabs of his own, more sure of himself than in the first round.

Few of the cadets here show natural ability (Haglin's opponent is an exception); but the worst appellation to earn in boxing isn't that of being a poor boxer—there are poor boxers by the double handful— but of being a coward. For that reason most of the plebes are willing to take a few punches just to get inside and land a solid blow.

Size is no indication of who will fight. A short cadet pulls his arms in tight, his fists beside his forehead, and glides in close to his opponent, coming in so low that his knee bangs the floor. All during this approach he is getting punched in the head; because he doesn't have much reach, the shorter fighter has to get in close to connect.

Forty-five minutes into the class Major White calls the plebes together; they stand, big gloves on their hips, breathing heavily. Their gray T-shirts are black with sweat at the collars and armpits.

"How do you feel?" White asks the class. They nod, grunt in some way that sounds vaguely positive.

"This is a game of chess in here. Only instead of losing your queen, you get punched in the head. You've got to think."

White moves about the ring, looking every cadet in the eye. Some of the plebes are a little disoriented. The next class, they know, they'll go up against the cadets in the other section, the ones in the next room who are learning with other instructors. They know the other men; they might be in the same classes, in the same company. They might even be roommates. But in the ring, the other boxer is just a big red target, a pair of gloves looking to connect with your nose.

"I'm going to tell you straight up," White says. "I don't want to go over there and lose."

He is low-key, his voice strong.

"Now, some of the other instructors are going to tell their people what to do for the first three or four punches. They're going to say, 'Go in there, throw two jabs, then a combination,' or 'Two jabs, then a straight right.' The plan is that if the boxer doesn't have to think, they'll get over that initial nervousness."

A couple of cadets nod when he mentions nerves.

"I'm going to make it even easier on you guys. I want you to walk across that ring, walk right up to your opponent and throw a big right cross. That's going to break the ice."

There are a few laughs in the crowd as they imagine how it might feel to take charge of the fight from the beginning.

"You're both going to be nervous; hit him first and you're going to pass all your nervousness over to your opponent. Now, a third of the guys I teach knock their opponents down when they do this."

The cadets in the audience chance a look around. One-third sounds like a big piece; but no one is about to dispute the towering figure in the ring. They're thinking about how to get through the next class, how to win a bout with a fighter they haven't yet laid eyes on. As always, the clock moves toward the next requirement.

At the end of the class Haglin removes his equipment; his face is red with perspiration and has deep lines from the tight-fitting head-gear. He watches his classmates stow their gloves and gear, makes sure the lockers are neat. Then the cadets form two files facing the door. Haglin calls the class to attention, salutes the instructor, turns, and dismisses the class. The sixteen men shout, "Beat the snot out of the other section, sir!" Then they hustle out of the room.

Haglin lives in Lee Barracks, the farthest from the Mess Hall, gym, and academic buildings. While many of the oldest barracks have been renovated and updated, Lee, built in 1962, is a bit shabby in compar- ison. Across the small courtyard, there is evidence of construction: a long plastic trash chute hangs from an open window into a dumpster. Because the rooms being renovated are empty, the cadets in Lee live three to a two-person room.

Haglin just lost his roommate, who left abruptly because of bad grades. On Saturday night Haglin was given twenty-four hours to move into a room with two other plebes. He moved everything, but the room is still in disarray and he isn't sure where some of his uniforms are. A sign on the door announces "PMI," (afternoon inspection) which is a more relaxed state of room repair than AMI (morning inspection) and SAMI (Saturday morning inspection—which requires the greatest preparation). The chain of command has allowed Haglin and his roommates some slack; they are allowed to have what is, by West Point standards, a messy room.

The L shaped desks, with space for computers, overwhelm the room. A set of bunk beds is pressed up against one wall; a tall dresser at the foot crowds the front of the wardrobe. The opposite wall holds a sink with two clothes hampers and medicine cabinets, a rifle rack, and coat closet. The closet, made for two sets of uniforms, is stuffed with three. On the floor beneath the beds are ranks of polished shoes and boots. The room has the rank smell peculiar to teenage boys: gym socks and athletic shoes and body odor.

On the top bunk, a blanket thrown like a coverlet seems startlingly out of place here. Haglin folds it, then goes into a desk drawer and pulls out a powdered sports drink, which he pours into a plastic bottle. He fills the bottle at the sink as he talks. He turns on the stereo, keeps the volume low. Study conditions. A Counting Crows song comes on, but just barely. Haglin talks about his Saturday night, which he spent moving his uniforms, books, computer, and issued gear from another floor. Still, the work didn't cut in on his social schedule.

"There's almost no social life here," he says. "It's a big deal if a couple of us get a pizza and hang out in someone's room to watch a movie [on a computer monitor]."

The plebes of this class are the first to have computers that will play movies from compact discs. No one from the class of 2001 on up misses the chance to comment that the plebes are going soft. The reality is three or four plebes, sitting on the cold floor, or on the hard desks or footlockers, trying to watch a feature movie on a computer screen.

On his bookshelf Haglin has a frame with a collage of photos. One shows a sign in front of his high school; the movable letters read,

"Pete Haglin: Good Luck at West Point." There are several shots of his family: one of him, his parents, and two sisters; another, a studio shot, shows him and his sisters, posing with their dog.

"I had a girlfriend at home," he says. "She sent me letters all through Beast, and packages of food. But then when I was at home [over Christmas leave] she started getting all serious and talking about our life together after college."

He shakes his head, takes a sip of his drink. "I can't imagine planning that far ahead right now."

Haglin's roommates come in. Cole is tall, with light brown hair, and a cool politeness. Berliner, the other roommate, has a boyish face and pleasant smile, curly hair, and small, round glasses. The three of them talk about an English paper due that day, then about the lunch menu (which they must memorize), then about Cole's move to another table in the Mess Hall.

"I was on some laid-back tables for a couple of months," Cole says. "I got used to it. Now I'm at a hard table. The other day I gave the table comm a glass that wasn't new, the plastic was a little cloudy. He hollered at me and said, 'That's a plebe cup,' and sent it back down."

Like all the plebes, these three have a finely honed sense of justice. They complain that the upper class take their food first, while the plebes get theirs last. No one at the table eats until the plebes have completed their duties, but the plebes notice when upperclass cadets take larger portions.

Haglin gathers his towel and excuses himself to go shower.

Berliner is from Bethel, Alaska; a town, he proudly reports, that you "can't drive to. You've got to fly or take a boat."

"I was surprised at how much people looked up to me because I was going to West Point," Berliner says. "Our town only ever sent one kid here before this, and he didn't make it through the first semester. When I was back home [at Christmas], they even asked me to be a guest speaker at this dinner in town," he says.

"But it brings some pressure," he admits. "Like you have to be on at all times."

Berliner ties a piece of flat rubber to one of the posts of the bunk bed, unfolds a paper that shows drawings of rehabilitative exercises, and begins moving his arms.

"I did this on the Indoor Obstacle Course," he reports, indicating his shoulder. "Wrenched it pulling myself up onto the shelf."

He managed to go on for a few more yards, but nearly lost his grip when he sidled out onto the bars ten feet above the gym floor. That's when he thought he should stop. Boxing didn't go much better for him.

"I'm famous for getting knocked out," he says unself-consciously.

Berliner got knocked down for an eight count, but got back up and resumed the fight.

"I found out later, after talking to the guy I boxed, that he was holding back. He was afraid he was going to hurt me." He smiles. "And I thought I was getting better."

Berliner landed a few punches, so whatever sympathy he had engendered disappeared. His opponent knocked him out. He tells this story while exercising his stick-thin arms.

Haglin returns from the locker room and begins to dress. He has a tattoo on his right shoulder, a squarish character. It is his Korean name: *Yung*, which means "Dragon."

"I got it over Columbus Day Weekend," he reports. "It's OK, as long as it doesn't show when you're in uniform."

The tattoo is one aspect of his "Preserve Pete Haglin" campaign. Body decorating provides a chance to express himself in a fashion that isn't designed by West Point, that isn't directed at some institutional goal. Haglin says the stocky boxer who pushed him around the ring this morning has a pierced tongue; incredibly, he wears the post all the time.

"If you really try, if you know it's there, you can see it when he talks," Haglin says. "He took it out during Beast, though."

Haglin and Berliner compare backgrounds. Both Army brats, they both lived in Korea as children. For Haglin, it was part homecoming, because he lived with his mother's parents for a while.

Cole, who has several military history books on his shelf, has no military experience in his family. His parents didn't want him to come to West Point.

"They didn't want me in harm's way," he says.

"My sister did well at Cornell. They were kind of hoping I'd follow her. They wanted to make sure I came here for the right reasons. Now they support me and are proud of me."

When Cole leaves, Berliner and Haglin exchange glances. Cole is a little too gung ho for their tastes. They point out the long line of boots and shoes under his bunk. Beside the footgear issued to cadets, there is a shiny pair of "jump boots," popular with paratroopers. They shine well but are no good in the field; they're associated with garrison, with polished floors and offices. They are not the muddy boots of "real soldiers." There is also a pair of highly shined "jungle boots," a Vietnam-era design with canvas uppers, made specifically for hot, wet conditions.

Berliner says Cole bought the boots with his own money even though they aren't needed.

"I want to be in the military and all," Haglin reports. "Being in the combat arms will be great. But I'm not going to go out and get a high and tight [buzz haircut] right now."

"I don't want to miss out completely on the college experience," he says, leaning over his bed and pulling the blanket taut.

"I want to be Pete Haglin as much as possible, and not just Cadet Haglin."

Later, in Grant Hall over an iced tea, Haglin admits to being a bit tired of the plebe game.

"I hate being everybody's monkey boy," he says. "We just look like such dorks walking around. 'Good morning sir! Good morning ma'am!' "

He also misses being able to choose his own friends. Haglin, who grew up on army posts, knows many upperclass cadets, other Army brats a year or two ahead of him. He swam with them, played ball, rode bikes. Now, they're off-limits because of rules against fraternization with the fourth class.

"The hazing is no big deal," he says. "I mean, at a fraternity people go through worse."

But he is tired of it: The move on Saturday into a room that was already full with two men; the fact that he can't talk to his friends; his "monkey-boy" status, fretting over whether or not an upperclass cadet gets a plastic glass that's a little too cloudy. Within these confines, Haglin looks for ways to assert himself.

"You control your own destiny," he says. "I told my squad leader that I wanted to work for a pride pass," he says. "Pride passes are like time off for good behavior."

Haglin's squad leader told him exactly what he would have to do, and Haglin succeeded. At a football game in the fall Haglin convinced some Rutgers University coeds to sit in the Army stands with the cadets.

" 'Good motivation, Haglin,' they told me."

He didn't need their praise, but he was happy to take it because it translated to a weekend pass.

"You tell them 'I want to earn x, they'll tell you what you need to do," he says approvingly. "My squad leader is concerned with my grades. He doesn't BS me just because I'm a plebe; he treats me like a person. Last thing in the world I want to do is get him in trouble because I'm not performing."

Haglin grudgingly admits that West Point has helped him in some ways. He is a better student and studies much more at the Academy than he would at another college. There is a bit of Epictetus about him, in the way he keeps his personal life, the life inside him, separate. Cadet Haglin spent Saturday night moving into an already crowded room with two cadets who would not be his first choice for roommates. Cadet Haglin doesn't like his haircut, or the "dorky" way he has to greet upperclass cadets.

Pete Haglin got a tattoo over a long weekend leave. Pete Haglin flirted with the girls from Rutgers and got them to come to the Army stands so that he could turn it into a "pride pass" and time off for good behavior.

He's winning some fights and losing some others. On balance, he says, "I'm pretty happy with myself."

"West Point is just something to get through," he says. "I feel bad about saying that, but it is . . . We live in this bubble. Some of these people just sit around on the weekends playing computer games; they don't try to get out and see other people."

As he sees it, this isn't helping create future leaders.

"No one is going to follow me to the grocery store unless I have a personality."

Haglin is trying to find a place for himself between different interpretations of what it means to be a cadet.

"I don't sit around and iron creases into my shirt," he says, fingering the fabric of his class uniform. Cole, his roommate, spends time each evening pressing his shirts, which come back from the laundry already pressed. Haglin's expression leaves no doubt as to what he thinks of this waste of time.

"Yet I've got more passes than either of my roommates."

He leans back in the chair, palms the empty drink bottle. All around, first and second class cadets use the small cafe tables to study. In the corner, a large-screen TV is tuned to some cable sports channel; two first class cadets stand in front of the set, thumbs hooked into their belts.

"Only twenty-nine and a butt days until Recognition," he says, finding his smile again.

Recognition is the ceremony that marks the end of the plebe year strictures. Plebes in each company line up like a wedding party while the upperclass cadets work their way down the line. The older cadets introduce themselves by their first names and, for the first time, call plebes by their first names. Plebes can then walk and talk like college students, or at least like upperclass cadets. They no longer have to march everywhere or formally greet upperclass cadets. They no longer have to act, in Haglin's words, like everybody's "monkey boy."

Plebes anticipate Recognition because it affects almost every aspect of their lives, but the fascination with time pervades cadet life.

They are forever counting the days until some event that promises to be better than whatever they're experiencing at the moment. They are young people whose strongest wish is that big chunks of their life would speed by them.

Haglin checks his watch, excuses himself. He needs to study chemistry, or the menu for lunch.

Over in Grant Barracks, Jacquelyn Messel and the other twenty-plus plebes of her company file into the dayroom after lunch. The basement room is clean and brightly painted, a long rectangle with an architectural anomaly: three huge posts, three feet square, march right down the middle of the room. The chairs and settees are arranged on either side, like a church with huge dividers running down the center aisle. Two televisions sit at the front of the room, one at the head of each set of benches. The plebes file in joylessly; they sit up straight, facing front. There are no feet on the furniture, no arms thrown over the backs of the chairs; no one turns around in his or her seat to chat with the classmate siting behind.

A third class cadet, the newly elected Honor Representative for the class of '01 in this company, begins. He looks sixteen.

"This class is about the NCO Honor Guide," he says, reading from some papers. He does not explain what this might be. Instead, he introduces Cadet Todd Morelli, the company First Sergeant, who strides to the front of the room. Morelli, who has dark hair and dark eyes, holds himself tightly and scowls from beneath his eyebrows.

"Duty, honor, country," he begins.

Morelli paces in the divided room, looking down one side, then striding briskly to the other. He recites the opening paragraphs of Douglas MacArthur's 1962 speech to the Corps of Cadets. This address gets a lot of play at West Point. It's part boosterism, part historical reminder, part cultural manifesto. It is an icon, as familiar to West Pointers as the Lord's Prayer is to Christians. But whatever else it is, it is not new. The plebes have not only heard it before; they are required to memorize parts of it.

Morelli goes through a couple of paragraphs. He doesn't tell the plebes why he's reading it. He moves from MacArthur, without comment, to a book in which a German officer tells of his treatment at the hands of his Russian captors during World War II.

The German prisoner was being abused by a guard; a Russian officer notices the German's Iron Cross—an award for valor—and stops the guard. In the story, the Russian officer points at the medal and tells the guard that the German is a soldier, a hero.

Morelli does not indicate if he is reading from a novel or a memoir.

"Honor, courage, loyalty," he says, "those are the things that will get you through. These will make people look up to you."

It is all he offers by way of explanation.

The plebes sit straight-backed in their chairs. If any of them wonder what Cadet Morelli is talking about, they don't show it in their body language. Morelli struts and preens in front of them, speaks in banal generalities.

Next Morelli reads the entire text of the Cadet Prayer.

At the end he looks up at his silent audience and, in a dramatic voice says, "I think that pretty much says it all. I can't say it better than that."

And, just in case some plebe is sitting there and wondering, *What does this have to do with me?*, Morelli adds, "If this doesn't mean anything to you, if you're just sitting here biding your time, you're in the wrong place."

With that warm lead-in, he asks for questions. There are none.

"I'm not here to push you away. I'm inviting you in," he scowls.

Jacque Messel pronounces the class "pretty good" without the least trace of sarcasm. On the way back to her room, she says that Morelli is famous throughout the corps. "We had a birthday party for Morelli; a whole regiment of plebes joined in."

Birthday parties are a way for plebes to strike back at unpopular upperclass cadets. Under the guise of a "spirit mission," they grab the offending cadet, douse him or her with shaving cream, soap, and shampoo. Then they hang the victim by feet and hands from one of the rolling laundry racks, like a lion brought back to the village.

Morelli knew what was coming.

"He had six friends waiting in his room to fight with the first plebes that rushed in," Messel says. "He climbed out the back window of the barracks and was trying to get away. We caught him though; there were too many of us for him to get away."

The plebes of First Regiment pushed the tied-up Morelli around the cadet area until the battalion commander, a firstie, broke up the fun. The story is an attempt at levity, but Messel resents Morelli's approach. The worst charge she levels against him isn't that he's hard on them; it's that he chooses his targets randomly.

"Somebody will be delivering laundry to his room and will disappear for an hour," she says.

One of the plebes who ventured near Morelli's room got "locked up," and had to stand at attention up against the wall.

"Morelli asked the plebe, 'Did you shine your brass?'" Messel says, referring to the brass belt buckle that needs attention every day. "The plebe answered 'Yes, sir,' and Morelli made the plebe show him the *back* of the buckle."

Predictably, the back was not shined.

"Morelli gave him a lecture on honor," Messel says, rolling her eyes. "On equivocation."

Morelli is the cadet first sergeant, the highest-ranking junior in the company. He came up with new duties for the plebes, duties that eat into their time and, the plebes complain, are unnecessary.

"Morelli has us cleaning and mopping the common areas in the company, wiping down the baseboards on stairwells. We have to clean the CQ [Charge of Quarters—a barracks orderly] area, the study room, and the dayroom."

Messel isn't complaining because she is afraid to get her hands dirty. The government pays civilian employees to clean the common areas so that cadets can use the time for their primary duty: study.

Morelli also dictates that the plebes do this work in the evening, just before study barracks. This negates the administration's plan to give cadets control over large blocks of time.

"You can't start studying or go work out. You can't get the work done ahead of time, you have to wait until that time to do it," Messel says. "It's just a lot of extra crap he thinks we should be doing now."

The duties, as she understands it, will not extend past Recognition Day. Instead of being a cause for celebration, she sees this as an affirmation that the duties are mere harassment.

Most of the problems plebes have, Messel says, depend on "who you mess up in front of."

"If you get on the First Sergeant's shit list, he'll follow you around and wait for you to mess up some more. He'll ask you more questions about your knowledge and give you more opportunities to mess up. You get higher visibility when what every plebe wants is lower visibility."

"I don't have a problem with hazing," she says. "I think it teaches you traditions. Upperclass cadets have to make corrections when they see things are wrong, but it shouldn't be random."

The first thing a visitor notices about Jacque Messel's room is that it smells a lot better than Pete Haglin's room. It isn't perfume; more shampoo, maybe. Like Haglin's room, it is also built for two, but is home to three. There is a bunk below the windows in what was meant to be open floor space; you have to step around it as soon as you enter. On the floor below the beds are the same shoes, clean and polished, lined up in a neat rank. This lineup includes black pumps with a modest heel, also shined to a furious sparkle, right beside the combat boots. One of the bookshelves holds a small teddy bear beside the neatly aligned books.

The fact that they can live this way astounds many of the plebes. These are the same American teenagers who, when they lived at home, left laundry piled up, mixed clean and dirty clothes in a wad at the bottom of the closet, who left shoes throughout the house. They left dishes in the sink and on the coffee table, returned the car with an empty gas tank.

There are changes that are highly visible, and other, more subtle differences.

"My parents treat me differently," Messel says, thinking of her visit home at Christmas.

"My mom and I were pretty good friends, but my dad and I used to fight about random things. Now I guess they see me as more of an adult. I'll call and say I'm going to New York City, and instead of giving me a bunch of instructions they say, 'Have a good time, tell so-and-so I said hi."

Like many college freshmen, Messel was a bit shocked by college work. High school wasn't hard for her, and she didn't study much. Instead, she devoted her time to an array of activities. Messel played volleyball, basketball, and ran track. She sang in a choir and was president of her school's National Honor Society, a finalist in the National Merit Scholarship competition. She didn't belong to any of the groups in her school, she says—the "scurvy" kids or the in-crowd with all the clothes and the drinking or the jocks or the band and choir. She had friends in each. She does admit, reluctantly, to being voted "Biggest Brain" in her senior class. She misses her friends, as well as the freedom she had.

Messel was a top candidate for admission. Because West Point was her first choice (and the only school she applied to) she received early notification: She learned in October of her senior year in high school that she'd been accepted. All of this seems ironic to her now, during the second semester of her plebe year, because her first semester average was a disappointing 2.75.

"I had a lot of trouble accepting that I wasn't going to be tops. I'm only going to be average here, so why try hard? I could spend an hour and a half editing a paper, or I could spend a fraction of that time to go to the gym or relax or hang out with my friends. I could even decide to go to bed so that I'll be able to stay awake in class the next day."

Messel's take on the constant time crunch reveals a problem, not just with the cadet schedule, but perhaps with Academy expectations. West Point demands that cadets do well in very many things:

Jacque Messel and Pete Haglin go from discussing poetry in English class to a roadmarch with full field pack during Unit Training Time in the after-school hours, then right back to studying. But the culture is not one that encourages cadets to excel in any one thing; instead, they are conditioned to handle multiple tasks. The result is an education that, some critics say, lacks depth. With so much on their plates, some cadets learn how to get by with minimum effort in many areas.

"It's hard," Messel says, "to balance all the requirements. And I miss being able to just sit and talk to people. I look for anything I can do to put this stuff out of my head for a while."

On a recent Friday night, Messel called a high school friend, now away at college, to wish the other woman a happy birthday. Her former buddies were just getting ready to go out, but Messel wasn't even near the end of her duties on a day that had started at 5:30. While she talked to them on the phone, she sat polishing her shoes and brass for inspection the next day.

"I'm getting the 'West Point attitude,'" she admits a little reluctantly. "That all college students are slackers. Mostly, I'm jealous."

This is a defensive reaction, similar to what old grads tell themselves to feel better about what they went through: *I had it tough and these kids today don't have it as bad.*

Cadets work hard and play hard, and they know it. It is a life that makes for hardworking, dedicated, aggressive soldiers. But there is a danger here, too, one that Jacque Messel acknowledges with her slightly embarrassed smile. It is elitism.

In an essay for an English class, one cadet lampooned this mind-set.

"They [cadets] love to talk about other colleges, where their old high school classmates 'party every night,' 'never go to class,' 'get all the girls,' 'slam heavy beers,' and live disgustingly hedonistic lives. The fantastic nature of this description of civilian life meets with no disagreement."

Another plebe, in an essay for English class, wrote about attending a concert at Eisenhower Hall, the cadet recreation center. He found himself "inspecting" the college students in the audience and silently criticizing their shoes ("gross"), haircuts ("what haircuts?"), dress, and even posture. Then in a flash of self-awareness, he thought, "What am I doing? Just a few months ago, that's what I looked like! What's happened to me?"

In *Making the Corps,* a dead-on study of the military and the society it serves, author Tom Ricks writes about young Marines going home on leave after basic training and being disgusted by the civilians around them, whom they suddenly see as unmotivated, slovenly, fat, lazy.

Pride and camaraderie are part of what makes the military work, but this separateness can lead to problems. West Point cadets, in particular, may be too isolated from their civilian peers. Many of the regular Army officers and NCOs at West Point blame this isolation for the fact that some cadets are ill-suited to relate to the young soldiers they will lead.

As strong sunlight comes through the big windows, warming the room, Jacque Messel sits at her desk, reviewing her daily schedule. She gets up at 5:30, wraps in a robe and shower shoes and makes her way to the female latrine, which is two floors down in this vertical barracks. She and her roommates clean their room: empty the wastebasket, dust every surface, polish the mirrors, clean the sink, put away all books and papers, ensure that all of the uniforms in the closet are in order (most formal to least formal) and hanging neatly. They sweep the floor, read the newspaper on-line for one or two articles they will have to command. They memorize the day's menu for breakfast (fried eggs, pancakes, maple syrup, orange juice, milk, yogurt, Canadian bacon), then head out to breakfast formation at 6:30.

They march to breakfast, do table duties ("There are four and a butt servings of Canadian bacon remaining on the table, sir!"), scarf down a meal that is notably short of fresh food, then hurry back to the barracks to pick up books. Classes and study in the mornings, with side trips to pick up uniforms at the tailors, shoes that have been resoled. There are appointments for shots and checkups, athletic equipment to draw or turn in. Lunch is a repeat of breakfast: the same accountability formation, the same menu of gray food. More classes in the afternoon, more studying, trying to get a start on group projects as she juggles daily assignments: physical education, math, chemistry, English, history. After class she hurries to dress for Team Handball. Then back to the barracks to clean the baseboards along the metal stairs, deliver laundry to upperclass rooms.

Messel goes to the Mess Hall to grab some take-out food for dinner and is in her room for the evening study period by 7:30. She and her roommates sit at their desks, help one another with common assignments. Sometimes they take a break, and Messel will play her guitar.

This schedule grinds on relentlessly for weeks at a time, which is why cadets—plebes especially—so look forward to the milestones that mark the changes in their status. For plebes, the biggest change comes with Recognition, when the most extreme strictures of plebe year end abruptly. Among the things that change with this ceremony: Upperclass men are able to talk to plebe women socially.

"The yearling and cow guys are paying more attention to the plebe women," Messel says. They are jockeying for position for when they are allowed to date, after graduation day and the end of the year. But for Messel, dating is too much of a distraction in the constant time crunch that is cadet life.

"It's not even close to my top priority," she says. Her needs are simpler than that right now. "I miss just being able to sit and talk to people. There's always something that needs to be done here."

During her first summer, Messel made comments like these as if she were justifying resigning. Now, she sounds like someone who has decided to cope.

"Beast Barracks was a tough time," she says. "But it's good to know you can make it through tough times. It gives you something you can use to compare other things with."

Messel enters Thayer Hall with the throngs in gray and finds her seat next to Pete Haglin. The class is a history survey course (American

History after 1865); the instructor is Major Rikard, a graduate of the University of Kentucky. Rikard is energetic, with a heavy accent and long vowels of his home state. Like most of the other instructors, Major Rikard's class is organized to an astonishing degree.

After taking the report, Rikard begins by going over an ID sheet the cadets completed in last class. Then he uses a slide to show a numerical analysis of how their section fared in comparison to other tests they've taken and to other sections' performance.

Every class meeting is organized around objectives and learning points; the study questions provide a framework, but they also demand that the instructor keep up a fairly rapid pace in order to cover the material. In a dozen other classrooms along this same hallway, other cadets are receiving the exact same instruction. The method ensures that all cadets in a particular course are graded on the same material according to the same standard. It does not provide much flexibility for either the students or the instructors.

Click. An outline of the class flashes on the board, down to the specific learning objectives and the topics for that lesson: "New Manifest Destiny," "A Splendid Little War," and "The New American Empire."

"What does imperialism mean to you?" Rikard asks, his voice powerful. He is a stocky man, with a gleaming bald head and glasses, and the body of a football player gone soft. He has a friendly face, the kind a stranger might approach to ask directions.

A cadet raises his hand. "It means that a nation expands its power through military force, economics, migration, and, uh, reverse assimilation."

Rikard raises his eyebrows, impressed.

"Good job, good job," he says encouragingly. "Glad to see you've been doing the readings."

The discussion moves on to Manifest Destiny. Along the back wall, Haglin looks sleepy. Messel jabs him with an elbow. Rikard puts up a slide that shows a nineteenth-century painting entitled *Manifest Destiny.* The painting shows an angelic Columbia, gilding over a landscape that looks like the Great Plains. Beneath and behind her, a railroad engine chugs west, as does a wagon train. She strings telegraph wire from one giant hand. Behind her is light, ahead of her is the darkness. In the lower left-hand corner, small dark figures—the first inhabitants of the Plains, scurry out of the way.

"God is telling us to expand west," a cadet explains.

Rikard puts up more images. One, an ad for soap, shows Admiral Dewey, hero of the Battle of Manila in America's war with Spain. Dewey, dressed in angelic white, is scrubbing his hands. The text talks about the white man's burden to bring civilization to the unwashed masses of the world. The symbols are unsophisticated, immediately apparent.

Given enough time, Rikard says, he might ask why Americans of a hundred years earlier, who considered themselves enlightened, fell for such obvious manipulation. He might pose a tougher question to his class: are Americans today still being manipulated? Will our advertising, our cultural artifacts look just as simplistic and misguided to an audience a hundred years from now? But there is too much to cover. The cadets scribble on their handout sheets, checking off the learning objectives.

Click. Another slide.

The discussion moves to the Spanish-American War. Rikard mentions that although popular opinion in the United States was inflamed by the "yellow press," the Spanish occupying Cuba were in fact brutal oppressors.

Click. Another slide.

Rikard says that the United States suffered 460 battle deaths, but lost 5,200 to disease.

The cadets write these numbers down. Across the room, Haglin is asleep. Rikard moves inside the horseshoe of desks to stand near him. When Haglin doesn't respond, the instructor tells him to stand.

Rikard moves quickly to another aspect of what he calls the "Spanish-American-Cuban-Filipino War."

The United States, he points out, quickly became bogged down in a guerrilla war, fighting Filipinos who didn't want to trade Spanish rulers for American rulers. Rikard mocks President McKinley's comments about annexing the Philippine Islands to "educate and uplift" the native people.

"It was about access to the markets in China," Rikard says.

Haglin, standing behind his chair, still looks sleepy. Beside him, Jacque Messel's eyelids keep sliding closed.

Rikard starts a film about the guerrilla war in the Philippines. There are black-and-white stills of American soldiers in a trench, firing their rifles. White smoke drifts over the scene. There are photos of twisted corpses in blue shirts and khaki pants; they are Americans. In another shot, the bodies are Filipino guerrillas.

"This was a brutal war," the film's narrator says.

Racism played a part in the brutality: It was in this conflict that the word "gook" was first used by GIs. The narrator points out that there was substantial resistance to the war on the home front. "Many saw a war against a native people seeking self-rule as a monstrous twist of American ideals."

The screen shows a photo of Andrew Carnegie, who offered to pay the United States government twenty million dollars to buy Philippine independence.

There is no sign that the cadets connect what's on the screen with their own future as soldiers in an imperial Army. They do not draw parallels between what happened in the Philippines and what happened in Vietnam. "We don't have much time for discussion," Rikard says. Since this is a survey course we just have so much material to cover that it's hard to go off on side conversations."

Critics of the West Point education—in uniform and out—say that it sacrifices depth for breadth. There is just too much going on. One graduate said that learning at West Point is "like trying to get a sip of water from an open fire hydrant." Defenders point out that a large number of West Pointers go on to post-graduate education, including one Rhodes Scholar in the Class of 1999 and four in the Class of 2000 (making the Academy fourth in the nation in number of Rhodes scholars).

West Point's education may or may not compare with the Ivy League, but it is certainly adequate to the task. Being a junior officer, a 1980 graduate said, isn't all that intellectually challenging.

"You don't have to be an Alvin Einstein," he said, thus proving his own point.

THE WAY HE SHOULD GO

Major Rob Olson walks quickly into his office, a sheaf of papers fluttering in his hand. Dressed in the long-sleeved shirt and black tie of his Class "A" uniform, he looks uncomfortable and—uncharacteristically—flustered.

"Well, I just got my ass handed to me by the garrison commander's staff," he says, smiling.

Olson is the executive officer for the next R-Day, which means that he and his boss are responsible for setting up the conditions necessary so that the cadet cadre, the legions of Tacs, medics, and civilian workers can turn these high schoolers and enlisted soldiers into cadets in the six hours between report time and the parade.

"This is my going-away present, Olson says. "I move [to Fort Leavenworth, Kansas] two days later."

The table in Olson's office is covered with folders, fanned out in neat columns and rows, all labeled with R-Day concerns: issue points, traffic control, parents' briefings. One sheet lists common parents' questions: "Will my son/daughter be able to try out for varsity sports over the summer?" and, "How do I get football tickets?"

R-Day is not a new event at West Point, of course, but the garrison commander—who controls most of the logistics, all of the military engineers, traffic, parking, everything that goes on at the post—is new to his job and wants to know the plan. His staff, eager to look good in the eyes of the new boss, asked Olson a lot of hardball questions.

"I used to be excited about the prospect of being in the toughest spot," he says. "I used to seek out the biggest challenges. But with this project," he waves a blue folder, "I'm having to force myself to pay attention. This is the first time I've ever had to do that, and I'm a little concerned."

He puts a dip of tobacco between his lips and jaw, fetches a soda can from the sink. He spits into the can. He is quieter.

"I'm a little concerned," he says again, sitting at the conference table in his office. "I hope I don't get out to the Army and find that I've lost it."

Outside the window, thin gray clouds skate by periodically. Everyone is watching the weather forecast with great interest, and the reports are still sketchy for graduation day. In the cadet area, things are in an uproar. The barracks are not in their usual state of upkeep. Firsties wander around in gym uniforms, moving stereos and computers and clothing and books, stacking the boxes in the hallways for shipment home or to their first duty station. Inside the rooms, bunks are unmade, every flat surface is covered with papers and books, the detritus of eight semesters of study, of forty-seven months of living in the barracks.

Outside the rooms of the first class, discarded cardboard boxes contain trash, reams of paper, worn-out three-ring binders, gray tunics and white shirts and even old shoes. There are footlockers, with fresh labels identifying their owners as second lieutenants.

The underclass rooms are better, but not by much. The plebes have packed up almost everything for the move out to Camp Buckner, where they will train after a short summer break. The yearlings and cows are scattering to assignments all over the Army, to leave, to schools at Fort Benning, to training all over the world. Duffel bags and suitcases and piles of uniforms compete with boxes for hallway floor space.

Sergeant First Class Donald Mercier, Company E-2's Tac NCO, is trying to keep account of who is going where, of last-minute changes to school assignments, last-minute additions to summer school. He's also trying to keep a handle on the barracks; he has to allow the cadets room and time to move, but he must also stress security of their belongings and safety above everything else.

Rob Olson is focused elsewhere.

"I've been reading a bunch of books from the business world to see how they define the difference between leadership and management," he says. "I know what I can't become, and that's just another manager, the guy who worries about a resource problem because it's a resource problem, instead of worrying about it because of the impact it would have on those who'll suffer because of it."

Olson is looking ahead to the assignment that will follow his ten months of schooling at the U.S. Army Command and General Staff College. When he joins an artillery unit in the field, Olson will be one of only two majors in the unit, the second or third most senior man in a five-hundred-man outfit. As executive officer of a battalion, he'll be squarely in charge of the staff officers and the unit's physical resources; he will manage resources and people so the commander is free to prepare the unit for war. The job is a necessary step, but Olson does not want to become another bean counter, sitting behind his desk looking at spreadsheets and numbers representing unit readiness.

But it isn't just his anticipation of new responsibilities that has him worried.

"My biggest fear is being promoted above my level of competence," he confesses. "I mean, I was a pretty high-speed artillery captain. What if I turn out to be the world's shittiest major?"

"It's not like when I was a lieutenant, and I was just one of a bunch of lieutenants. I'll be one of two majors in the whole unit. I'll go in there and people will probably know I'm a BZ [early promotion, called "below the zone"]; they'll expect a lot. I don't want to be in a situation where somebody pulls the curtain back and there's the wizard . . . and I'm him."

Olson believes his fear will be enough to keep him honest. The right level of anxiety will help him pay close attention to his new responsibilities. He also knows that leaving this job will be hard on him. "I'm leaving something I really love," he says.

The evening before, the company had a "Hail & Farewell," a social event that is a tradition throughout the army: Arriving soldiers and families are welcomed, and the group says good-bye to those moving on.

"First, we welcomed the new Tac and said good bye to the yearlings."

Cadets are shuffled to new companies at the end of their second year, the whole class redistributed throughout the corps.

"Then the firsties. Saying good bye to them, that was the toughest speech I've ever had to give. Holly even said to me, after I was finished, 'You got a little choked-up there. Glad to see you've got at least one sentimental bone in your body.' "

Olson spits into the can, then leans back in the chair and surveys the office, the walls covered with guidons, unit flags, plaques, and mementos, including his West Point diploma and his master's degree. In the corner, the drawers of a gray, government-issue file cabinet are marked with class years: 1999 at the top, 2002 at the bottom.

"Usually by this time, this close to leaving, I'd already have most of my shit packed up and ready to move."

Nothing is packed. There isn't even an empty cardboard box waiting to be filled.

Olson has never been as closely attached to the development of young people as he has been as a Tac. He likens the experience to watching his daughter head to the bus on her first day of school. The connection isn't a function of how much time he's spent with cadets. In fact, in Regular Army units he has spent more time close to soldiers, but there was always the gulf of rank. Most soldiers are not trying to emulate their officers, but cadets do watch their Tacs and say "I want to be like that," or "I don't want to be like that."

Olson is constantly amazed at how much he has learned from his job, especially when it comes to communicating. For instance, second class cadets have the same level of intelligence as a second lieutenant, but have absolutely no reference point of experience to help them make sense of things. It was up to Olson to give them the "why" with the "what."

"You can just say, 'Pay attention to detail,' but that isn't going to sink in. For a kid on the jump team, I relate it to a well-packed parachute. For a kid on the mountaineering team, it's tying knots. For a football player, it's blocking assignments; for a kid who's prior-service, you talk about other things. I'm so much better at communicating because I've had to do those things."

This came as something of a revelation to Olson.

"[When] they invited me back here to get an MA, and I was petrified, given the pain associated with getting that bachelor's degree. Four years of being told I'm not the sharpest knife in the drawer, and then I'm going to check in for more schooling and get my ass whupped at a whole new level? But I found I could do these things."

The most obvious result of Olson's confidence was that he was willing to let the cadets take charge, willing to let them learn from experience. Not all Tacs have bought into this approach, and that is understandable. It is the Tac, after all, who is answerable to his chain of command. And while cadet mistakes may cost them reprimands or room confinement, all those marks go away on graduation day, like a juvenile record that is sealed when the offender turns eighteen. An officer's performance, in the form of annual fitness reports, stays with him or her.

"But letting the cadets see what goes on, by letting them run the system, that extra 10 percent of insight they get in how things work—they're going to be way ahead. In fact, I'd put any of my firsties up against Rob Shaw [the first captain]."

Olson is criticized by peers who say he doesn't have to worry about his career because he's married to a doctor; his kids aren't going to starve. He knows that people whisper that he can afford to be daring because he was selected for early promotion and further schooling even before he started work as a Tac. But the question for him isn't whether or not Rob Olson would have operated differently if he were the only breadwinner in his family, or whether it would have made a difference if he were still in competition for a slot at Command and General Staff College. The question that guides Olson is: How does his style affect cadets? Does it make them better leaders?

The first drill period of the day on Wednesday is at 7:15. It is cool enough so that several cadets, sitting in short sleeves in the Superintendent's box, have goose bumps on their arms. The chain of command practices, then heads back into the barracks as the rest of the corps filters out for the full rehearsal. Firsties in class uniforms wearing tarbucket hats, sabers, and white gloves mill about as the underclassmen spill out of the barracks. In spite of the hour, in spite of the fact that many of them were out late the night before, they joke around and enjoy their last hours together in this group.

Around the parade field, the first families to arrive for the week are making their appearances. One middle-aged couple is decked out in full souvenir regalia: the man is wearing a West Point T-shirt and hat; he has a camera bag around his neck. The woman has on a sweatshirt that says "West Point Mom." She carries a cloth bag with a full-color Academy crest. They smile happily.

The corps is finally assembled. The sky is mostly sun, although a few clouds skid by. The cadets are in class uniform: short-sleeved gray shirt, gray trousers with black stripe. They wear the uncomfortable tarbucket parade hats. The under class carry rifles, the firsties carry sabers.

Nearly four thousand cadets march out to the final line. They execute a series of commands: saluting with their rifles, coming to attention, to parade rest, to present arms and back again, practicing the sequence. There are now second class cadets in charge, thrust into roles they aren't used to, moving the entire corps across the parade field.

Colonel Joe Adamczyk, the Brigade Tactical Officer, strides across the parade field and takes the microphone from the cadet announcer. He tells the cadets that they don't look sharp, that they'll be out here until they get it right. There are some groans in the ranks, but the cadets, especially the seniors, want the parade to look good for their families.

Round and round they go. The companies wheel off, one by one, into a long line that snakes east, then two left turns, then close by the reviewing stand where the Superintendent's party will stand, and toward Quarters 100. Instead of going back into the barracks, they return to their starting positions.

The cadets still look happy and relaxed, even the plebes. The Class of 2002 has just returned from two days at Camp Buckner, where they moved in, cleaned the barracks, set up their field gear. For them, the year is already over, and they are, mentally at least, already on summer leave. For the yearlings, most of whom are headed out to take a long look at the real army, there is the promise of adventure. For the firsties, there is the almost unbelievable realization that this is their week, that the ceremonies are finally for them. Only the second class seem to have no fun as they struggle with their new roles running the corps.

After they march forward, the graduating class stands in loose ranks just in front of the bleachers. They remove their tarbucket hats, showing hair pressed with perspiration. A few of them already look exhausted from the week's celebrations, the last late nights in the barracks.

Kevin Bradley stands with F-2, his black hat under one arm, his saber belt hanging loosely on him. He looks a little bleary-eyed after a late night with friends. On the way back to the barracks, one of them announced, "Well, that's one day of graduation week over." They rode quietly the rest of the way. Although much of a cadet's life is about anticipation—a four-year recitation of "The Days"—the firsties still don't seem prepared for this. The entire week is a mixture of elation and melancholy and the more pedestrian concerns of playing host. Bradley's parents are due in around 11:00.

"I'm shipping them off on the tour of Constitution Island," he says, smiling and pointing at the historic site in the river.

"My mom has been waiting to see that for four years; I thought we should get her over there. Then we have the Supe's reception this afternoon; that's when things are going to start getting crazy."

But the week has already gotten off to a rough start.

Cadet Chad Jones was Bradley's roommate first semester by nature of their jobs. Jones was the commander of the Second Regiment, one of the five highest-ranking cadets in the corps; Bradley was his executive officer, the second-in-command. Jones, who was also class president, was outgoing, friendly, and confident. Part of that was probably a function of his age; he had served as an enlisted soldier before attending the USMA Prep School and is several years older than most of his classmates.

Jones had branched infantry. During the graduation ceremony, he was to present a gift, on behalf of the class, to the guest speaker. He was to be given his diploma by the guest speaker, along with the honor graduates and the first captain. Jones was due to be married a few months after graduation; Kevin was planning on going to the wedding.

At the beginning of graduation week, the Superintendent, Lieutenant General Christman, separated Chad Jones from the Corps of Cadets for cheating on a history paper.

Jones used the entire bibliography from another cadet's paper, as well as the same quotations. But he did not document the work or indicate that it wasn't his own, and the professor caught the plagiarism. Jones, brought in front of an Honor Board in the late spring, denied that he had done anything wrong. His defense was that he had been sloppy in his scholarship; he should have noted his sources.

The investigations lasted for several weeks. When the hearings finally started, Bradley attended. "I saw the people on that board," he says. "And I knew they were trying to do the right thing. But they also felt a lot of pressure to make an example of him."

Jones was hardly a struggling student; Bradley thinks his GPA was about a 3.2. "He almost always had his papers done ahead of time," Bradley says. "I don't know what could have happened."

The class officers removed Jones as president.

Behind Bradley, the parade announcers read the names and hometowns of the second class cadets who now command the formations passing by. It's a travel map of America. Aspen, Colorado; Idaho Falls, Idaho; San Antonio, Texas; Bridgeton, New Jersey; Malibu, California; Pleasant Garden, North Carolina; Elberton, Georgia; Pelican Rapids, Minnesota.

Later, Bradley sits in Grant Hall. Families are gathering here, with their West Point guidebooks, cameras, and walking shoes. They study the life-sized paintings of MacArthur, Marshall, Eisenhower. An elderly couple, speaking Japanese, has their photo taken in front of MacArthur. The general is not smiling in the picture.

"This was a lose-lose situation," Bradley says of Jones's case. "If the Supe gave him discretion, everyone would say, 'Oh, as long as you're a regimental commander you can do pretty much anything and still graduate.' If the Supe throws him out, people will say, 'But he gave discretion to all those other people. They're still here.' "

In Bradley's own company, a member of the class of 2000 was caught using a fake ID card while at Fort Benning, Georgia. "That's not something you do accidentally," Bradley says.

Yet Christman used his discretion. The cadet, found by an Honor Board, left the Academy during the first semester and will rejoin the class of 2001. He was suspended for less than a year, sent back to the next class.

The Superintendent's wide use of discretion has saved some cadets who will no doubt be fine officers. In fact, it could well be that these men and women will be the best officers, having learned their lesson about honor the hard way.

But all of this has happened at a cost to the corps and to the

Honor Code. As Colonel Peter Stromberg of the English Department said, Christman has "hoisted himself on his own petard." His wide use of discretion has led some cadets to think, not that there are cases that must be decided individually, but that some lies aren't as bad as others.

"It's made the Honor Code something of a joke," Bradley says. "It breeds cynicism because people look around and say 'worse people than him have stayed.' "

Another aspect of the problem, as Bradley sees it, is that the use of discretion will vary from Superintendent to Superintendent. His memory of General Graves, Christman's predecessor, is that more cadets were separated. And whether or not this is true doesn't really matter; the corps believes it's true.

Bradley feels that part of the problem with the Honor Code is that most cadets think of it as an institution, rather than a set of values, as something that exists—or does not exist—inside each of them. "We don't have a lot of interaction with the Honor Committee. They could be living right down the hall and you wouldn't know it unless they put a little sign on the door. How do we know they represent us?"

But the aspect of the Jones case that weighs heavily on his mind is the non-toleration clause. Discussions of the incident in his company, among his classmates, have focused on that again and again.

"It's not a hard question," Bradley says, "unless it's a close friend."

The honor system calls for someone who believes another has violated the code to confront that cadet. Bradley says he would have confronted Jones immediately if he had known of the plagiarism. The tough part comes if the accused cadet refuses to own up; the other cadet is bound to report the incident to the Honor Board. This critical aspect of the code is contained in its last words, "or tolerate those who do."

"I don't know what I would have done in that situation," he says, shaking his head. "That's always a question on emerging leader boards."

"Emerging leader boards" are interviews with cadets being considered for high-ranking positions.

"I told my board that it was the right thing [to turn someone in] and that I hoped I could do the right thing."

Bradley is miserable. He is everything West Point tries to produce, a thinking man, a person of character, dedicated to his profession, thoughtful and articulate and selfless. The epitome of the Academy product.

"I wouldn't be able to turn in my best friend," he says at last.

In Central Area, cadets in BDUs haul heavy duffel bags to waiting trucks. A sign at the head of one line reads "PCS." The PCS line is for graduating seniors, who will do a permanent change of station. The other sign reads "TDY," for temporary duty. This line is for underclass cadets headed to Army posts around the world for summer training.

The cadets walking through Central Area give each other elaborate handshakes. They hug and call out duty stations and promises to stay in touch. For the under class, this is better than a football Saturday. For the seniors, it is also stressful.

Kris Yagel says he wants to spend time with his friends, and he feels a little guilty because he's thinking of his family as another responsibility to juggle. The Academy grounds are not designed for entertaining; the cadets are not used to having dozens, even scores of visitors to drag around post.

Still, West Point puts on its best face for graduation week, and the formal garden beside the Superintendent's quarters is clipped and swept and pruned to perfection for the reception. Roses bloom along the ornate brick wall. There is a wooden guardhouse, about the size of an old-fashioned telephone booth, painted white and green. During the nineteenth century, cadets used it to stand guard during summer encampments on the Plain. The formal garden is a showcase for plants and flowering shrubs donated by parents' clubs from around the country: azaleas from California, climbing hydrangea from Michigan, tulips from Arkansas, grape holly from Oregon, bluebells from Virginia.

The receptions are organized by battalion; the first class cadets from three companies at a time are to show up with their families at the appointed hour. With twelve battalions in the corps, it is a long two days for the Superintendent and Susan Christman, who greet every single guest in a receiving line.

Military social gatherings are peculiar in that everyone shows up at the hour specified on the invitation. In Army units, it is customary for the commander to have the officers and spouses to visit. If the invitation says fourteen hundred, all the couples are on the front steps of the commander's quarters at a few minutes before two. These may be the only parties in the world where there are fifty guests and only one knock at the door.

True to form, cadets in their starched white uniforms gather in front of Quarters 100 at the appointed hour, waiting for their families who must contend with the shortage of parking. Many of the cadets look impatient. One young man scans the distant sidewalks for his family, doing everything but tapping his foot.

"They're running late," he tells me. "I told them it would be tough to park."

Kevin Bradley looks a little nervous, too. It isn't that anyone is checking off names at the gate, but a command performance is a command performance.

Bradley once stood up his family in a similar circumstance. During his plebe parent weekend, he was to meet his parents for a reception. He went to his room, got dressed and, noticing that he had a few minutes to spare, did what any cadet would do: he took a nap. Forty-five minutes after he was supposed to have met his family, a cadet was shaking him awake.

"Dad was really pissed," he says now, smiling at the memory.

Dave and Marge Bradley make it on time, and they line up with the other families. At a table set up on the lawn, women from the Officers' Wives Club are selling West Point memorabilia and homemade crafts: Christmas tree ornaments made to look like cadets, note paper and postcards and cookbooks and pins made out of cadet buttons.

At the head of the receiving line, a cheerful sign in gold with black lettering instructs the cadets on the proper way to negotiate the receiving line. (They've had this lesson before, in etiquette classes.)

The sign tells the cadet how to arrange his or her family: mother and father first, then grandparents, aunts, uncles, siblings. No helpful hints for firsties who show up with mom and dad, and mom's new husband and dad's new wife. They're supposed to be able to think on their feet by this time.

Most firsties have been in the Supe's house at least one time (there was a reception earlier in the year for first class cadets). Most of the parents have only seen the home from the parade field. Some are clearly in awe of the surroundings.

The Christmans stand under a leafy arbor. The Supe's aide, at the head of the line, learns the first names of each guest from the cadet and passes it along. Dan Christman gives each family the same big smile he passes around so generously to cadets, the same hearty handshake. (He has removed his class ring for these trials, so that his fingers don't get crushed.) Then Christman leans over and passes the name to his wife, Susan, who smiles charmingly and offers a few pleasantries, just as she will do for thousands of other guests.

"Welcome to West Point, you must be so proud. This is a great day for your family."

Just off the sunroom where the Christmans eat lunch is Sylvanus Thayer's sundial, ordered from London and installed here in 1831. In the 1850s, when Robert E. Lee was Superintendent, the grounds had a pond and a boathouse.

Under a white tent fly, the families gather and eat cookies and drink punch. There are officers from the staff and faculty here, along with their spouses. It is one of the social duties that come with high-ranking positions at West Point.

Ann Stromberg, whose husband Peter is the senior member of the academic board, confides that this is must be her "ninety-seventh appearance" at one of these.

"And the only really dreadful ones were when they served Jello punch."

A full colonel introduces himself to the Bradleys after noticing that their son wears armor branch insignia. The colonel was an armor officer before becoming a professor. His wife tells stories of moving about from post to post, of their favorite assignments and best quarters. The couple doesn't talk about the moves in terms of endless uprootings. And although the newspapers are filled with speculation about intervention and a possible ground war in eastern Europe, no one mentions Kosovo.

After a half hour of chatting, it is time to go (also specified on the invitation). Bradley tells his parents he'd like to have Chad Jones join them for dinner at a local restaurant. The Bradleys think that's a good idea.

The seniors are everywhere, floating on the excitement, glittering in the adoring eyes of their families. Every cadet is a hero this week: All sins are forgiven. In their white tunics they look like priests; this is their season, their week, their moment.

On the grassy field across from the library, behind General Patton's statue, formations of second class cadets in BDUs practice with sabers, learning how to draw them, how to salute with them, how to return them to the scabbard, all without taking off an ear or a finger. In a few short days, they will be seniors, and it will be their job to run the corps.

In Thayer Hall, the cadet gospel choir gives a concert. The huge auditorium is not close to being filled, but the families there to watch make up for their small numbers with sheer enthusiasm. Colonel Maureen LeBoeuf sits down front with her young daughter, the two of them whispering and enjoying themselves and clapping to the music. Colonel Adamczyk and his wife sit a few rows back. He does not keep time with the music, but he is here, supporting the cadets and making another public appearance in a week of social hyperactivity.

Johnny Goff, who quarterbacked Army's football team last season, is one of the featured singers. He has a good voice, higher than one might expect from a Division I football player. When he is finished, he stands at the microphone and quotes from Proverbs. "Train a child in the way he should go, and when he is old he will not turn from it."

Then he looks out into the audience and asks his parents to stand so he can thank them. They do, and the families applaud wildly, finding an outlet for all the emotion welling up.

While the first class enjoys the festivities at West Point, the class of 2002 moves some thirteen miles from main post to Camp Buckner, where they will begin summer training after their long-awaited three-week leave. A large plaque beneath the flagpole tells visitors that the camp is named for Lieutenant General Simon Bolivar Buckner, USMA 1908, who was killed in action on Okinawa, June 18, 1945.

The new residents have cleaned the barracks, policed the grounds, stowed their uniforms and equipment, met their new chain of command, found their way to the snack bar and Mess Hall. In the morning, they will load trucks at 0445 to return to West Point for the rest of the week's parades. But for this evening at least, they have an abundance of that most precious commodity: free time. This is genuine free time, not an hour stolen from the Dean with thoughts of a paper that's due soon, not a slice wrestled from studying or athletics or sleep. This is the real thing, the genuine article. They cannot leave the camp, but they are finished with their duties and are practically giddy with relief as plebe year draws to a close and summer leave heaves into view.

The Mess Hall staff has set up grills outside, cooking hamburgers and hot dogs, the smoke curling around the trees and many small buildings.

Camp Buckner straddles the western end of West Point's reservation, tucked hard alongside a lake amid low-slung hills. Buckner is a small town, with its own barracks, infirmary, theater, mess hall, guardhouse, and supply facilities; this is where the new yearlings spend their summer, learning about the branches of the army.

On this Wednesday evening before Saturday's graduation, the sky is clear, with a promise of stars. It is too early in the season for the swarms of gnats and flies that will plague them for two months, nor is there any taste of the crushing humidity that will fill the barracks for the rest of the summer. Right now, the cadets are dressed comfortably in PT uniforms, laughing and joking and simply stunned that this day has finally come.

Like almost all of her classmates this evening, Jacque Messel wears an oversize USMA sweatshirt over her PT uniform of T-shirt and black shorts. Her hair is pulled back in a short ponytail. She is relaxed and smiling as she walks along the road that passes between the barracks, mostly metal buildings about eighty feet long by twenty feet wide, tossed haphazardly on the wooded hillside. Around her, the thousand-plus members of the class of 2002 play Frisbee or basketball or just sit on the ground, talking under the trees.

"My grades were good," Messel says. Then, a moment later, this former "Brainiest Girl" in her high school class says, "Not where I'd like them to be, though."

She has to pick a major in the fall. She had planned on chemistry or some other science that would prepare her for medical school; she has adjusted her sights.

"Even the chem majors who validated plebe chemistry [tested out of the course] are up until 4:00 A.M. studying."

"The longer I stay here the more reasons I see that make me want to stay," she says. There is no longer anything tentative about her; this is not a trial run. "You miss things that people get to do at other colleges—but you get stuff here that they don't: the camaraderie, the shared experiences, stuff like the Army-Navy game. Everyone takes the same classes, does the same training, so you never really go through anything alone."

This is the essence of the cadet experience: The group is everything.

She passes the small theater; there will be a show that evening, a hypnotist. Every few yards, someone calls out to her.

"Hey Jacque, what company are you in?"

The class is re-organized into different companies for the summer, which gives the cadets a chance to know and work with even more of their classmates. Messel is looking forward to the adventures, perhaps because getting here hasn't been easy.

"You can lose it if you just focus on what other people [outside of West Point] are doing."

A plebe in her company left just after exams, sticking around only long enough to earn credit for the semester. "He had gone to a military high school, to USMAPS [the Military Academy Prep School]. He quit because he couldn't be himself. You can get yourself into that mental state if you just think about things like: 'They tell me when to eat, when to sleep, what to wear.' But you also get a chance to be with friends, and even get away."

"After a month of Beast, I had no idea who I was or what I was doing. By the academic year it started to look a little more like a college with military training thrown in. There are some people who let themselves be defined by West Point, then you take that away and they're lost."

That reminds her of a story about cadets who can't get away.

"A bunch of us went to New York City last weekend, after exams, with the guy who was leaving our company. We went in this place and half the people there were cadets; you could just tell," she says, laughing.

On a manicured athletic field beside the road, a dozen cadets are playing "ultimate Frisbee." Just to the right is the Confidence Obstacle Course Jacque and her classmates will negotiate this summer, with its forty-foot tower and its fifteen-foot drop into the lake.

"There was this one guy, he had on a BDU belt with his civilian clothes."

The government-issue black web belt is both functional and ugly. But the regulation says a cadet can't leave without a belt. Since this cadet apparently didn't own a civilian belt, he wore his BDU belt.

"A chick magnet," and she laughs again.

This is what she must have been like a month or so before R-Day, when she was enjoying her high school graduation and her friends.

"The biggest compliment we got when we were in Mexico [on Spring Break] was when some guy would say, 'You guys are cadets?' We were paranoid. We'd look around and say to each other, 'Look, there are *real* college girls.' Of course, when the guys found out we were West Point cadets, they wanted to know how many push-ups we could do."

The highlight of her year was joining team handball. She grew close to her teammates and got to travel away from West Point. Team handball looks a little like indoor soccer, a little like basketball. A tall former volleyball player like Messel is a good fit.

She wasn't surprised to find that sports turned out to be a saving factor. During Beast the previous summer, second class Greg Stitt, her platoon sergeant, talked to her about sports and activities to help her imagine what life would be like in the academic year. He helped her see past basic training, to put it in perspective.

Stitt had a pizza party after second-semester exams for the plebes of his former platoon. Messel still talks about him with some awe in her voice; he took the time to figure out how to communicate with her.

If team handball was the highlight of her year, the low point came in the spring, when a first class cadet on her table was reported missing at taps one night.

"Everyone who knew him knew right away what had happened," she says, lowering her voice.

Cadet Eric Roderick's car turned up abandoned in the parking lot of a small restaurant five miles south of West Point, a half mile north of where the Bear Mountain Bridge spans the Hudson. There was no suicide note, no indication that he was depressed or upset. In fact, the senior, twenty-two years old and just over a month from graduation, had just been accepted at medical school at Ohio University.

"I sat at his table," Messel says. "He was pretty cool. He had his stuff together. He worked out hard, he studied hard, he was so psyched when he got into medical school. He could crack a joke, and he would talk to you like you were a person."

Messel had a chance to talk to Roderick after she was picked to serve on an Honor Board. Since no plebe in her company had pulled this duty, she asked Roderick about it. When she told him she was worried that her questions were stupid, he put her mind at ease and said, "No, just ask."

It is still hard for her to reconcile that this young man she saw every day is gone. Harder still because his motivation is so difficult for her to understand. Eric Roderick was a thrill-seeker.

"When he was in high school he used to jump off bridges and into mineshafts. He used to brag about it to people."

"There is every indication that this was a thrill-seeking activity that went bad," Captain John Cornelio of West Point's public information office told a local newspaper. "He had mentioned to classmates that he wanted to jump off the bridge."

The roadbed is over 160 feet above the water. Roderick's body was not found.

Messel describes the midnight "Taps Vigil," the corps' traditional way of saying good-bye to one of its members.

At the end of the evening study period, the corps assembles on the concrete apron beside the parade field, and the lights are all out in the barracks. The four thousand cadets are silent. There is just the shuffle of leather-soled shoes on concrete, the sound of the barracks doors opening and closing. In front of them, out in the darkness along diagonal walk, a bugler plays taps, then a bagpiper follows with "Amazing Grace." Finally, the corps sings the alma mater.

> *And when our work is done*
> *Our course on earth is run,*
> *May it be said, "Well Done!"*
> *Be thou at peace*

Then the cadets go back to their rooms. Messel tells the story as she walks to the end of the camp and some picnic tables placed beside the lake. The evening is cool, and she pulls her knees to her chest as she sits on the bench.

She is looking forward to leave, to seeing her friends and her family, to revisiting one of her family's summer spots at the Lake of the

Ozarks. Boating, skiing, swimming, and "touristy" stuff; the simple pleasures of being able to drive again, to sleep in a little, to decide her own schedule. Pete Haglin had planned to come and visit her over the summer. She saw a lot of Haglin during the spring semester, she says, "because he had a crush on my roommate." But Haglin will be in summer school.

"He landed in STAP by .3 percent," she says. He'd let stuff go too long, and then he'd work really hard to try to catch up. But he didn't make it with chem. He always talks about quitting; I know he was really looking forward to getting out of here this summer. But he'll be back."

And so will Jacque Messel. West Point is a different place for her than it was ten months earlier. She and her classmates have finished the intense period of learning how to follow. Now they begin the climb to leadership positions.

"Since Recognition we've gotten more and more responsibility," Messel says. "And that'll increase once we're out here [at Camp Buckner]. We'll rotate through as team leaders; one of my classmates is acting squad leader right now because the squad leader isn't here. It's scary and exciting."

Whatever lessons Jacque Messel has learned about leadership—good and bad—have come from observing others. She feels lucky to have had, during the academic year, good team leaders. Each had strong qualities she wants to emulate.

"Dan Young was my team leader first semester; he always stuck up for me. I had friends whose team leaders were afraid to approach the squad leader, the platoon sergeant, the platoon leader. This spring my name came down on this list saying I needed to make up a road march [from Beast]. But I knew I had done enough to qualify."

Including the last few hundred meters of the march back from Lake Frederick, going up the ski slope. Her team leader explained this to everyone in the cadet chain of command. All they said was "Your name is on the list, you gotta do the road march."

The march presented no great physical challenge. It would be an inconvenience more than anything. But the list was wrong. Messel didn't want to go along with the program just to excuse bad record-keeping, and she had every right to be taken at her word.

"My team leader went to the Tac. I think she was impressed he would do that. She said, 'If you're qualified, you're qualified.' "

At the same time, her team leader did not try to shield her from trouble when she messed up, or did not know her plebe knowledge. She is glad he took this approach, since it is important to Messel that she stand on her own feet. Many of the cadet men get a "big-brother" mind-set and want to do everything for the women, but Messel knew that approach would come back to haunt her. So she did things on her own.

"Being a woman here isn't as bad as I thought it would be. I thought we'd be isolated, but it doesn't happen much, mostly when the guys are afraid. Sometimes they worry too much about offending us. I figure if you offend easily you're in the wrong place anyway."

As Messel heads back to the barracks, a three-quarter moon rolls along the treetops. Along the road leading away from the camp, cadet guards walk their post. They wear BDUs, pistol belts with canteens, and they carry flashlights. A van pulls up and three cadets, second class women in starched whites, climb out, still laughing about some story they'd been telling. An officer passing by chats them up; they have returned from a dinner for the women's crew team.

This is a place of groups and subgroups. A cadet belongs to a team and to a certain platoon in a particular company, to a class, to a group of friends. Belonging is the defining feature of cadet life. This is how the whole is held together, by the crosshatched and overlapped connections that run like tape wrapped around a package.

Bob Friesema emerges from Pershing Barracks two days before graduation, wearing the white over gray summer dress uniform. He looks like an upperclassman, though he still wears the plebe's brass "U.S." on the shoulders of his dress white shirt. He is relaxed, confident, and nothing like he was during Beast. He handled every challenge thrown

at him, most very well; yet he had a wild-eyed, nervous look throughout Beast.

Nine months later, his hair has grown to the point where he can
comb it, and he no longer moves as if he's had too much coffee. He'll
be a good model for the admissions candidates he will meet at home
over summer leave.

"I wanted to know the small details," he says. "That's what I'll talk
to them about. Like the fact that my family addressed all my mail to
'Cadet Friesema' instead of to 'New Cadet Friesema.' And every time
I got a letter, some upper class would say, 'So, you think you've
already made it though Beast, huh?' I did a few push ups for that."

Still, he says, life was relatively easy during Beast. "You get yelled
at, you do some push-ups. But you didn't get hours [on the area] or
Article 10s [administrative punishment]. Of course the crisis times
were my fault. I'd procrastinate on some assignment and get behind.
That was stressful, as opposed to scary. I came into the semester
thinking that if I just stayed out of summer school I'd be happy. I try
not to think like that any more."

Like Messel, Friesema went to the pizza party Greg Stitt had for
his former charges at the end of the year. He arrived late, but he stuck
around after the others left to talk to Stitt. The widely admired junior
talked to Friesema about when to do a West Point detail [as Beast or
Buckner cadre], about how to pick a major. He also told Friesema to
keep his military grade as high as possible because that grade affects
all a cadet's assignments.

"Afterwards I talked to him about his perspective on Beast,
because I knew they [the cadre] saw it differently. They were up later
than we were and got up earlier. They were accountable to the Tacs
and the chain of command; we were only accountable to one person,
the squad leader."

The prospect of increased responsibility doesn't make him nervous; it's what he most looks forward to.

"You have to know that tough times are necessary, and it'll be better for you in the long run."

Then he quotes one of Stitt's favorite sayings: "Pain is just weak-

ness leaving the body." In a year or two, some other class of plebes
will be talking about Friesema and quoting it as his.

His experience with Stitt, Grady Jett, and Shannon Stein, with his
own team leaders from the academic year, has shaped what Bob
Friesema thinks about leadership. Not one of the plebes mentioned a
lesson in leadership from an academic setting, from a classroom. Few
of the plebes mentioned any officers (from whom they are separated
by layers of upperclass cadets). Their whole experience in this most
formative year came from other cadets. In particular, from the cadets
they were closest to. Not the Kevin Bradleys, up there in the rarefied
air of company command, but the team leaders and squad leaders
they saw every day.

"My first detail team leader was from Wisconsin [Friesema's
home state]. All the plebes hated him because he was tough and
could be mean. He wasn't that way to me. He used to ask me what
kind of music I liked, and he'd have that on when I came around to
his room for some duty."

Even though he was the benefactor of this small kindness,
Friesema thought the favoritism was wrong.

"You shouldn't have favorites or different ways of treating people.
He did look out for me, though. He made sure I was squared away
and knew my stuff so that I didn't get in trouble with the squad
leader or platoon sergeant. He used to ask me if I was getting enough
to eat, not because he wanted to be my buddy, but because he knew
that plebes sometimes don't get enough. Sometimes he'd have these
little boxes of cereal in his room, from the Mess Hall, and he'd give
me some."

His second-detail team leader was more laid-back and interested
in Friesema's intellectual development. He had Friesema read *A
Clockwork Orange, Lord of the Flies,* and *Blackhawk Down* (an
account of the disastrous American military operation in Mogadishu,
Somali in 1993), then they'd discuss the books. Although he enjoyed
the dialogue, he also recognized that the second-semester team
leader was extremely cynical about West Point, about the endless
rules and requirements.

Friesema tried to take a longer perspective. "This stuff just gets old if people don't know why they're doing something. I mean, some things you hate wouldn't be so bad if you knew why you were doing them. I'm sure someone somewhere thinks there's a reason for it."

Friesema has lunch in Grant Hall, where the little snack bar is preparing for the onslaught of graduation week visitors. They're building sandwiches assembly-line fashion, stacking them in shiny plastic boxes in the cooler. Afterwards he walks around Trophy Point, with its displays of cannons captured in the nation's wars. At the foot of the flagpole, a couple of dozen pieces lie on steel rails. They range from a few feet long to fifteen-foot monsters; some are shiny with fresh black paint. Sunlight drifts downstream on the river's surface. The weather is perfect, with just enough clouds to show a contrast with the blue sky. The hills are touched with green, the air light and dry and still, so that the view up the river is unobstructed.

There are graduating firsties everywhere, escorting visitors, strolling around the Plain. They wear a uniform peculiar to this week: white shirt over gray trousers, red sash, no hat. They are all smiling, taking their time. Friesema's walk is leisurely, too. Like Jacque Messel, he seems happy here. Part of the reason things have been looking up for him is that he started seeing a young woman who lives in nearby Connecticut. He met her through a friend and plans to invite her to the formal dance at Camp Buckner during the summer.

There are other reasons for his optimism: he'll soon be on leave, and he'll spend the time at home with his family. He is also looking to the next set of challenges the Academy will throw at him.

"I can't wait to meet my plebe in the fall," he says.

Early on Friday morning a thousand first class cadets drag themselves up the steep hill to Michie Stadium for graduation practice. At 8:00 the sun is already bright, promising a hot day. A lieutenant colonel stands at the microphone, speaking rapidly about the sequence of events for the graduation ceremony.

"Do not move beyond this line until the graduates ahead of you have reached the bottom of the ramp," he says, pointing to the speaker's dais. "Do not let your sabers bang against the seats. Turn to the center of the stage when you get your diploma. If the guest of honor wants to shake your hand, do not walk by him."

The litany goes on, and the firsties slump in their seats. Captain Gillian Boice, Jacque Messel's Tac, suggests that more than a few of them are hung over from celebrating the night before.

Colonel Adamczyk, the brigade tactical officer, takes the microphone and reminds them to be careful.

"Don't do anything stupid that would cause you to miss your graduation. Look out for each other tonight, take care of each other."

He does not tell them, "Don't drink and drive." They've heard that admonition plenty of times. Instead, he talks about consequences. The prize, so close, can still be yanked away. He also mentions that he'll be in the back row of dignitaries on the stage, "So your chances of hitting me with a hat are pretty slim." The joke draws only a few chuckles.

Down in the cadet area, the end of one year merely means a transition to the next. In Grant Hall, five juniors sit at a long table with a major; each has a file folder holding plans for the summer. The officer points out something on one of the sheets, the cadets follow along.

In the basement of Washington Hall, in the uniform shop, a stack of winter dress coats sits on the floor. The tailors will add a stripe to each sleeve, indicating that the coat's owner has moved up a class.

A line of firsties waits outside another room. They hold pillows and sheets and telephones and saber belts: all government-issue material which must be returned before they can leave post. With twenty-five hours to go until graduation, they are standing in line, which is probably what they were doing twenty-five hours after becoming new cadets.

The graduation parade is scheduled for 10:00 Friday morning, which is earlier than in the past. Someone in the Commandant's office did a little research and found that most spring thunderstorms at West

Point occur in the late afternoon, which was when the graduation parade was traditionally held.

At Captain Brian Turner's graduation parade in 1991, dark clouds rolled in and drenched cadets and spectators. The downfall was so intense that some seniors broke ranks and slid along the flooded grass. They wore their summer whites for the next day's graduation ceremony, after a heroic effort by the cadet laundry to dry-clean all one thousand dress coats failed.

Today there is no threat of rain; the sunlight is already strong and hot as the stands begin to fill around nine. One family uses the pages from the post newspaper, folded into paper hats, for shade. Officers in heavy dress greens—most are on escort duty—cluster in the shade of some trees behind the reviewing stands.

The president of Cambodia is here because his son is graduating. A dozen U.S. Secret Service agents, looking like movie typecasts with their dark glasses, boring suits, and curly wires in their ears, stand around behind the Superintendent's reviewing stand. They look very relaxed; one of them smokes a cigarette as he stands by a dark sedan.

Not every cadet is marching. Lynn Haseman and Meghann Sullivan, Shannon Stein's roommates from first semester, got off all-night guard just before the parade. They are excused from marching, but want to watch because they never see the ceremony from this end.

"This is really cool," Haseman admits.

Sullivan's hair is dyed a pretty blond. The two young women look healthy, a little tanned, eager for their summer assignments. When Stein's name comes up, Sullivan shakes her head, finds something to look at on the field.

"Shannon got in a little trouble," Haseman says. "She's going to get an award this afternoon [her varsity letter], then she goes to see the Tac to get punished."

Stein is accused of fraternization with a plebe man.

"Shannon is a real type-A personality," Haseman says. "She doesn't take criticism well, so you really can't tell her much. She has some issues she needs to work out. She makes some bad decisions, but no one wants to confront her because she'll attack."

Haseman knows what she's talking about: She also made some bad choices first semester. She had come back from a bar and gone to a plebe's room to "bum a smoke." A little bit of flirting, a little bit of being where she wasn't supposed to be, and she was also accused of fraternization.

"It was no big deal," she begins. "But I can see why the Tac thought it was a big deal. I decided I had to make better decisions."

The Graduation Parade, the last parade for the class of 1999, takes place the day before graduation. The ceremony is the reverse of the Acceptance Day Parade, which was their first as cadets. In that parade, at the end of Beast Barracks, the new class stands near the bleachers. When the rest of the corps assembles on the Plain, the new plebes leave the reviewing line and join their new companies. Four years later, the same cadets march onto the parade field with the rest of the corps. Then, on command, they leave their companies and march to the reviewing line, the same spot they occupied four years earlier.

The stands are packed. Most of the people in the press section give up their seats to senior citizens. A little thrill shivers through the crowd as ten cadets march briskly out onto the field, stopping at what look like random points in the huge space (there are markers on the ground, like stage marks, that are invisible from the stands).

These are the adjutants. The rest of the corps will emerge and use them as guides. A bugle spits out four or five sharp notes: "Adjutant's Call." Then in a moment, the entire band starts playing, and the music rolls across the bright grass, sending little shivers through the families gathered in the bleachers and standing five deep on the sidewalks around the field.

The companies come out of the sally ports in solid blocks of gray over white; hands and arms swing just so, legs bend at the same angles. The sun cracks into splinters on bayonets, sabers, shiny brass breastplates.

The reviewing line, just in front of the stands, is marked with metal replicas of the company guidons. Every few feet, another pole, another gray and gold-metal pennant, starting with Company A-1 on the far left, to Company H-4 on the far right. Many of the families have positioned themselves behind the guidon where their son or daughter will stand. After the corps has gone through the ritual salutes, after the singing of the National Anthem, the first class is ready to leave.

"For-ward . . ."

The commands echo down the long line.

"March!"

They approach the crowd in a long rank. As they get close, people in the stands call out to them; the firsties are only yards from the spectators.

The Class of 1999 reaches the mark, then, company by company, they execute a sharp about-face and remove the big tarbuckets. They hold their hats in the crook of the left arm, like a trophy. The pose, the plumed hats and shiny brass, looks like something from the eighteenth century.

The three classes left on the field salute the graduating seniors, then prepare to pass in review. Instead of large staffs at the head of each unit, there is now a single second class cadet.

Lynn Haseman and Meghann Sullivan begin a joking, *sotto voce* commentary on their classmates who lead the formations.

"He's so cute," Haseman says of one. "What a great guy." Then, "What a loser."

In the strong sunlight, the tight cotton pants the cadets wear are nearly transparent.

"Looking good in your boxers," Haseman says to another classmate. Only Sullivan can hear her.

The graduating class stands in a long rank just in front of the bleachers. The rest of the corps marches in front of them. Each company of firsties puts their hats back on and renders a hand salute as their own unit passes by. Shoulders back, heads up, they watch as the next class takes charge.

The corps, reduced by nearly a quarter, disappears back into the sally ports; the band plays "The Army Song," which ends all ceremonies here. The firsties look a little surprised to find themselves still on the field after it's over.

A voice on the loudspeaker asks spectators to please keep off the Plain, a request that is immediately ignored as family members pour across the chain to find sons and daughters, to form little bands for pictures: now with the grandparents, now with a brother, now with Mom and Dad. One cadet protests when his mother asks him to put his tarbucket back on for a photo, but most of the graduating seniors take all the attention in good humor.

Dave and Madge Bradley make it to Trophy Point just as Kevin does. He pulls the gleaming saber from the cloth sash, hands it to his mother, tells them it is a gift for them. (It is not the government-issue saber he has worn all year, but one he bought from the same supplier.)

Madge doesn't quite know what to do with it. She kisses her son on the cheek, cradles the shiny saber in her arm the way a pageant winner might hold a bouquet of roses. She is overwhelmed by it all.

Dave Bradley is a bundle of excess energy. He paces and looks for Kevin's brother and sister and great aunt, who are driving up from New Jersey this morning. He worries about where they will park, if they got lost, if they will hold up Bradley, who has another assembly to attend. Madge spends her nervous energy talking, telling stories about her brothers and children. She moves the saber from one arm to the other.

Kevin Bradley is a bit overwhelmed, too, or perhaps just exhausted. There has been no time for him to rest, to take a break. Even as his body was engaged in the most mundane tasks—standing in line to turn in a telephone—he has been very much aware that he will soon leave his friends.

There is another family posing for pictures at this high point, with the river in the background. A sailboat appears below, its sail on fire with sunlight, as if put there by some set director to make the scene even more perfect. The cadet's mother orchestrates the photo session; she is beside herself with emotion.

"You must be the mom," she says, hugging Madge. "I can tell by how proud you look."

"My son gave me a saber at Christmas," the woman confides. "It's nice to know that after all these years, they finally appreciate what you've done for them.

Cori and Madge talk about the dance that night: dinner in the Mess Hall followed by a formal at Eisenhower Hall. Cori, who is fifteen, has a new dress. Madge says she paid particular attention to her shoes, so that she wouldn't be taller than her big brother when she danced with him. Bradley's brother, Sean, will be at the hotel to meet the rest of the group coming in for graduation: relatives and friends and some of Bradley's coaches from high school.

The Bradleys take a few photos. Kevin tells his parents that he needs to be at Eisenhower Hall in an hour to receive his academic award; he will graduate number twenty in his class. Dave Bradley, still energized, and now with his other son, daughter, and great aunt in tow, organizes and briefs everyone. They'll meet in front of the Thayer Monument after Bradley changes uniforms yet again. Dave pushes his aunt's wheelchair; the family threads its way through the crowds taking pictures on Battle Monument. Kevin is in some quiet zone, as if trying to memorize every moment. He goes ahead, into the line of cadets headed back to the barracks.

That evening the firsties fill the Mess Hall, a last dinner for the Class of 1999, after thousands of meals here. The hangar-sized space, dark as mid-winter, is transformed by the company: by the new suits and gowns, the white uniforms, the smiles and glitter and excitement.

After dinner, the crowd streams out into the warm evening air, past MacArthur's statue, past the Superintendent's house, down to Eisenhower Hall and the ballroom with its twenty-foot-high windows that overlook the river. The evening is clear, the lights of Newburgh and Cold Spring reflect on the water. Tonight it all seems part of a romantic adventure, an elaborate stage set of youth and beauty. Tonight, everything is possible, and the great adventure awaits them. For a moment beneath the lights and the music, alongside this river, the graduating seniors reap their rewards. They have paid their dues.

GRADUATION DAY

No one is glittering at 5:15 on Saturday morning as the cadets spill out of their barracks for formation. Major Rob Olson, however, is already in full form.

"Here we go," he says in a voice that's much too loud for the hour. "They went to bed at zero two, up at zero-four. Oh, yeah, and tonight they're gonna drive all night to see that girl in Ohio."

He steps in and around the cars that pack Central Area. The firsties have been loading their belongings. Most of the cars and trucks and sport utility vehicles are filled to bursting.

"Somewhere around western Pennsylvania that Motel Six is gonna start looking real good," Olson says.

The sleepy cadets are startled by the sound of his voice, which carries loudly through the area. He may be the only person speaking. Olson keeps up his running commentary, teasing cadets as they stumble forward to find their companies.

"Looking real good there," he says to one, who salutes without looking up. "You sleep under a rock?"

Out on the Plain, where Second Regiment is forming up, the grass is covered with a thick morning fog. The cadets in white over gray look ghostlike as they wade through the mist.

This morning's ceremony is a promotion for the under classes. The second, third, and fourth class cadets wear the insignia of their new rank on their epaulets; the epaulets are turned upside down. During the ceremony, they'll spin the shoulder insignia over, so that the new rank shows.

Captain Brian Turner is here, his arms crossed against the morning chill. He smells of mouthwash and soap; his uniform, as always, is perfectly pressed. He salutes Olson, gives a perfunctory greeting. Olson goes off to join his own company.

Firstie Nick Albrecht, one of F-2's cheering section for the Sandhurst Competition, is in class uniform in the rear of the formation. He wears a name-tag and the black shield of a senior, but no insignia of rank and no branch insignia. This is the same configuration the recently dismissed Chad Jones wore, the almost uniform of those on their way out.

"He's graduating, but maybe he shouldn't," Turner says angrily.

He says this for Albrecht's benefit. The cadet, who is only a few feet away, pretends not to hear. "He can't even get the uniform right. A bunch of these guys got carried away and threw away their white over gray. Then they say, 'Sir, nobody *told* me to keep my white over gray.' " He imitates a childish whine as he says this.

"That's a kid's answer. I ask them if they think they can pull that kind of stuff in the Army."

There are six or seven cadets—firsties—in the wrong uniform, dark gray shirts and overseas caps. They stand out in the formation of white hats and white shirts. Albrecht holds a deflated football.

"It's kind of an award," he says. He explains that it's a tradition in F-2, for the first class to give the football to the second class cadet who can hold his drink better than anyone else. Albrecht holds up the partially inflated ball and shows off the names and dates, which go back to the nineteen sixties.

He leaves the story unfinished.

"Go ahead," Turner prompts. Albrecht doesn't respond.

"I wouldn't let them have a drinking party," Turner says.

Around them, the thousand cadets of the regiment are formed in a giant horseshoe. In each company, bleary-eyed firsties promote the second class, then move down the ranks to the third and finally, the fourth class.

Albrecht and Turner pointedly ignore each other.

Turner knows that command-sanctioned drinking parties are an anathema in the Army; one DUI conviction will end a career. Not to mention the fact that soldiers expect officers—even twenty-two-year-old lieutenants—to comport themselves like adults, not like fraternity boys. But the cadets know that Turner doesn't drink alcohol; they figure that his decision is personal.

"The cadets told me I was one of the toughest Tacs in the corps," Turner says. "Which is just a polite way of calling me an asshole. They thought I was too much in their stuff, that I was getting too involved."

Arms folded, he looks out over the formation. There is no joking around at this hour.

"Maybe. Sometimes that's part of being in charge. You know, when I was a cadet it used to be: 'Do this because I said so.' I tried to stop that. I tell them why they're doing something so they'll understand when they get to the army. Some get it, some don't. But they can't say they never heard it."

Turner is disappointed that only one firstie, Kevin Bradley, took him up on an offer to "come by the office to talk about the army." This especially troubles him because one of his chief complaints about his own superiors is that they do not make themselves available to mentor younger officers.

Turner's overall level of frustration also reflects a sense of alienation from his peers. Brigadier General Ron Johnson, Turner's former math instructor and something of a mentor for the younger officer, said Brian felt left out when other Tacs went to play golf or go fishing, two things Turner doesn't do. Johnson told him to take a day off and learn.

"Tacs aren't given enough latitude, enough decision-making

authority," Turner says. "There's too much due process. Tacs are almost afraid to give a low military development grade. It's not a matter of, 'What's the right thing to do?' Instead, it becomes. 'I don't want to embarrass my boss.' "

Turner and the other Tacs are required to defend the grades they give cadets, and the Brigade Tactical Officer, Colonel Adamczyk, asks tough questions about low scores. Turner knows this because he just recommended a failing military development grade—and dismissal from West Point—for a second class cadet.

"He was on the overweight program, he hadn't passed his Indoor Obstacle Course or the APFT [Army Physical Fitness Test]."

By second class year, the government has invested a great deal of money in each cadet; they are not dismissed without a lot of scrutiny. Was everything done to help the cadet meet the standard? Did the cadet understand the seriousness of his situation? Turner didn't see the scrutiny as part of a check, a nod to due process. He takes it as an indictment of his judgment.

He mentions his own upcoming fitness report, his report card for the entire year. Officers are evaluated on their performance; they are also compared to their peers in the same job. Since Tacs are carefully chosen from among the most successful officers in the army, Turner is competing in a fast group.

"I'm pretty down about it," he says, "because I don't really know where I stand. I guess all I can say is that I've given 100 percent every day since I became an officer. Guess that means nothing. I get tired of going through this."

By graduation week, Turner has decided to make a change. He told the Army that he wants to leave his primary specialty as an armor officer and work in the Acquisition Corps, which negotiates contracts to supply the Army's vast needs.

"I know I have what it takes to be [an armor] battalion commander," he says. "I'm not sure I'd ever get the chance, though, and I don't want to wait around all those years to find out."

Yet he doesn't know much about the Acquisition Corps, and confesses that he signed up for it because it sounded like something he could use later in life.

"I can still change my mind about it," he says, twice.

The pressures of being a single officer on an isolated post, of being a minority, of competing in a field with some of the best officers of his rank, are piling up on him. He needs a chance to unwind, to get away from West Point and get some perspective; there is an opportunity in the three weeks before he begins work as a Beast Tac. He plans to take a long weekend.

The fog is lifting off the field. Turner stands aside as the cadets march past, then up the front steps of Washington Hall and through the big Mess Hall doors for another twenty-minute meal.

For four years the class of 1999 has been counting the days until graduation. It has always been far in the future. Finally, shockingly, unbelievably, it is here.

The steep road leading to Michie Stadium is filled with underclass cadets in white over gray. The platoons tramp behind company guidons and sing marching songs.

Used to date a beauty queen
Now I got my M16

Just below the south end of the stadium, in the shade of some trees by the beautiful alumni center, a white tent fly is set up for the families of F-2's graduating cadets. Later in the day there will be a pinning ceremony; the new lieutenants, changed into the dress uniforms of officers, will have their gold bars pinned on by family members.

On the football field, the combined choirs and glee club sit on low bleachers beside the USMA band. The VIPs onstage include a US Senator and several members of Congress; the Secretary of the Army, Luis Caldera, USMA '78; and General Dennis Reimer, '62, the soon-to-retire Chief of Staff of the Army. Many of the families clustered in the stands learned from their experience during yesterday's sun-drenched parade: there are a few parasols in view.

The stage is set up in the eastern end zone, and long metal benches for the graduating cadets face the dais. Behind the seniors sit the tactical officers and NCOs. With the exception of the honor grad-uates, who occupy in the first row, the cadets are arranged by com-pany. Beside the ramps leading off the stage are four members of the Class of 1949, which has unofficially adopted the Class of 1999. Three retired generals and a colonel (the class's most-decorated combat vet-eran) will hand each graduate a small box containing the gold bars of a second lieutenant, inscribed on the back "USMA '49–'99."

On a bench in the press section sits a First Sergeant who wears the shoulder patch of the Second Infantry Division. The former Tac NCO is on leave from his assignment in Korea and has come up here to see "his" cadets graduate. In front of him, the Tacs sweat in their heavy uniforms; they tap their black shoes as the band plays.

Captain Gillian Boice, Jacque Messel's Tac, is looking for some tissues, just in case she cries.

"I tell my cadets that I'm strong, but I'm passionate."

Everything happens quickly now. The class marches in to applause from the stands. Their sabers clatter as they settle, tightly packed on the metal benches. When they sit, they remove their white hats. The Superintendent goes to the microphone; he invites the cadets to stand and applaud their families, to thank them for all their support. The cadets do so enthusiastically.

Christman then mentions that he was a plebe at General Reimer's graduation in 1962. One memory of that day stands out clearly, he says, almost forty years later. He pauses for comic timing. "I was thinking, 'Plebe year is over and I'm going home on leave.' "

The graduates laugh politely.

Christman also remembers the speaker. President John F. Kennedy promised the cadets that the world was changing and the future would hold many challenges for them.

"Kennedy was right," Christman says. "And the same holds true for you."

It is the kind of pronouncement they've been hearing for four years: standard graduation-speech maxims. With all these families present, it would be impolite for Christman to point out that many of the cadets present that day in 1962 would die in Vietnam within a few years of their commander-in-chief's speech.

General Dennis Reimer is the keynote speaker. This will be his last official function at West Point; he retires in two weeks, after thirty-seven years in uniform. Reimer starts off with a funny story about soldiers, then his talk drags on into a loose collection of quotations from other people's speeches. The audience drifts, and the families in the stands fidget. He is, mercifully, brief.

The combined choirs stand to sing "The Corps." The cadets and graduates, who memorized the song as part of plebe-year knowledge, join in, singing about the Long Gray Line.

The Corps, bareheaded salute it
With eyes up thanking our God
That we of the Corps are treading
Where they of the Corps have trod

The Dean of the Academic Board, Brigadier General Fletcher Lamkin, takes the microphone. "These graduating cadets have been awarded the degree of Bachelor of Science," he says.

The first two rows of cadets stand; these are the honor graduates. Walt Cooper, the number-one man in order of merit, is ranked first in his class. He will attend the U.S. Army Ranger School—an intense small-unit tactics course—over the summer. In the fall he will move to England, where he will study as a Rhodes Scholar.

Kevin Bradley, who is number twenty in order of merit, receives his diploma from General Reimer, shakes hands with Secretary Caldera, then comes off the stage smiling. After the honor graduates, the newly elected class president and the first captain receive their diplomas.

Then two readers and two presenters line up at the head of the twin ramps; they read the names more rapidly. Officials have taken great care to get the correct pronunciation of every cadet's name;

each white tube has the typed name and the phonetic spelling. The litany of names is punctuated with wild cheering as families in the stands acknowledge their cadets. Some of the cadets are over-whelmed as they come off the stage. They dance, clasp friends in tight embraces. Many of them are on the verge of tears, and quite a few fall to their knees to pray.

As the graduates of E-2 come off the stage, Major Rob Olson and Sergeant First Class Mercier position themselves alongside the ramp. Olson, brimming with emotion, hugs each of his charges.

Suddenly the class breaks into wild cheering for the class "goat," the cadet ranked dead-last in order of merit. She receives her diploma with one hand; in the other she holds a brown paper bag containing a dollar bill from each of 937 classmates. It is, tradition holds, her reward for keeping the bottom of the class from falling out.

After the presentation of diplomas, the Corps of Cadets is invited to sing the Alma Mater. The seniors, the underclass cadets, and old grads in the stands sing along. The song is unabashedly emotional, a holdover from a romantic era; but for these few minutes, at least, all cynicism seems banished, and every man and woman is a believer in duty, honor, country.

> *Hail Alma Mater dear,*
> *To us be ever near*
> *Help us thy motto bear, through all the years*
> *Let duty be well performed*
> *Honor be e'er untarn'd*
> *Country be ever armed*
> *West Point, by thee*

The tune is soft and slow, almost a lullaby.

> *Guide us they sons, aright*
> *Teach us by day, by night*
> *To keep thine honor bright*
> *For thee to fight*

When we depart from thee
Serving on land or sea
May we still loyal be
West Point to thee

And when our work is done
Our course on earth is run
May it be said, "Well done!"
Be thou at peace

The music swells at the end; even the least musically inclined are singing now.

E'er may that line of gray
Increase from day to day
Live, serve and die we pray,
West Point for thee

After the alma mater, the popular Commandant, Brigadier General John Abizaid, walks to the microphone to administer the oath of office. The 937 men and women, their heads uncovered, raise their right hands.

I . . . having been appointed an officer in the United States Army in the grade of second lieutenant, do solemnly swear that I will support and defend the Constitution of the United States against all enemies, foreign and domestic; that I will bear true faith and allegiance to the same; that I take this obligation freely without any mental reservation or purpose of evasion; and that I will well and faithfully discharge the duties of the office upon which I am about to enter, so help me God.

Inside two or three minutes they have ended one journey— symbolized by the diplomas inside the white tubes—and have embarked on another with a solemn oath.

Up in the bleachers, the rest of the corps stands to watch the closing moments of the ceremony. It moves each of them a little closer to their own day. In the end zone behind the graduates, hundreds of children mill around behind the linked hands of cadet ushers. On the field, photographers position themselves to capture the moment that has come to symbolize graduation here: the hat toss. When the class is dismissed, the new lieutenants throw their white hats into the air (they no longer need the cadet cap). Children are allowed on the field to retrieve a souvenir hat. Cadets will sometimes put photographs, notes, even money inside the caps. And because this is a place of rules, there is a paragraph inside today's program that spells out the procedure: Children must be between six and twelve years old, and "for the safety of the youngsters," between thirty-six and fifty-four inches tall.

"In order to provide maximum opportunity for all, children will be strictly limited to one hat. . . . If you are unable to locate your child," the brochure goes on without giving any hint of the melee that will ensue, "the Military Police will assist you in locating them at Gate 3."

In photographs of graduation, the hats are always shown at the top of the arc. But that is just an illusion.

There is no stasis, of course. *Tempus fugit* and all that. Even as the first captain is being called forward to dismiss the class, hundreds of the younger cadets are thinking about their next duty, about their flight overseas, about summer school classes, about leave and Camp Buckner. Down in the cadet area, government-issue equipment is put away to make room for the shipments of new shoes and boots and blankets and uniforms for the Class of 2003. The Tacs ponder who among the class of 2000 is ready to handle the responsibilities the summer will bring. West Point is about forward motion and jam-packed days; these are men and women of action, above all else. There is no hesitation, almost no time for reflection.

Robert Shaw, former first captain and now a second lieutenant of infantry, comes forward and centers himself before the speaker's platform. He gives the command for his classmates to "Re . . . cover," at which they all put their white hats on for the last time.

General Christman tells Shaw to take charge and dismiss the class.

Shaw turns and bellows, "Class of 1999, dis-missed!"

Instantly the caps sail into the bright blue. And perhaps the cycle does hold still, if only for an instant, if only for as long as it takes the white hats to climb, pause at the top of the arc, tumble over in the sunlight, and fall back to earth.

EPILOGUE: SUMMER 2000

Rob and Holly Olson got their joint assignment to Fort Leavenworth, then Holly spent the last months of their tour attending a career course for medical officers at Fort Sam Houston, Texas. She commuted to Kansas most weekends. The Olsons put up with this relatively minor inconvenience so that they could move together to their next assignment in Hawaii, where Rob was to join the 25th Infantry Division for a second tour. Within weeks of arriving, Olson and his new unit deployed back to the mainland for a month of combat training.

Brian Turner became more disillusioned with his chain of command during his second year as a Tac, 1999–2000. "I never thought I'd say I hate coming to work in the morning. Working with these kids is great, but I don't respect the senior officers I work for." Turner applied to the Federal Bureau of Investigation, because he still wants to serve his country. As of summer, 2000, he is waiting for the FBI to lift a hiring freeze.

Within a year after graduation, **Kevin Bradley** was on patrol out of Camp Monteith, Kosovo. "We set up checkpoints, search cars, provide escorts, and things of that nature," he says. "The most surprising thing to me and the thing I was not prepared for was the magnitude of the responsibility the PLs [platoon leaders, a lieutenant's job] have. Each is assigned a certain number of towns. . . . They have to meet with the mayors and the town councils to listen to their problems (especially difficult in towns with both Serbs and Albanians) and keep the peace. It is a pretty brutal schedule . . . twenty-four hours a day, seven days a week."

When it came time to choose his branch, **Grady Jett** picked the Field Artillery, which has been popular among football players in recent years. He also selected, for his first assignment, Fort Hood, in his home state of Texas. Jett plans on leaving the Army after his five-year commitment is up, although he says he is going in with an open mind. "I may stay in if I'm enjoying the unit and the people. If I'm not enjoying it, there's no reason to stay in."

Shannon Stein branched Military Intelligence, but tried to change when an officer told her that, at her first assignment, she'd probably spend her days "getting coffee for colonels." Stein will "definitely" leave the service after her five-year commitment. Among her friends, she "cannot think of a single person who wants to make a career" of the Army. "I can't see myself staying in. No one has really impressed me that much—I mean, I've met some great officers, but I don't want to be like them."

Greg Stitt married classmate Sarah Hatton on the day after graduation for the Class of 2000. The newlyweds were both headed to Fort Rucker, Alabama for aviation training. Stitt, who enlisted in the Army to become a pilot, is pursuing his dream.

Alisha Bryan spent much of senior year recuperating from reconstructive knee surgery. She devoted her free time to planning a

Minority Youth Conference, where high school students and their cadet hosts talked about the virtues of education, personal awareness, and responsibility. Bryan selected the Quartermaster Corps and a first assignment in Germany, and is looking forward to her mother visiting her overseas.

Bob Friesema concentrated on academics his yearling year and was rewarded with his highest GPA ever. He enjoyed his introduction to the Army's branches during training at Camp Buckner, and by the spring of 2000 had volunteered for parachute training. He also spent part of his third summer on duty with a U.S. Armored Cavalry unit in Germany. The prospect of returning for cow year and the long commitment to the Army did not give him pause. "It's not even a question for me."

Jacque Messel's yearling year was marked by athletic triumphs, including a trip to national level competition with team handball. In the spring of her second year she was a little surprised to find herself preparing to be a Beast squad leader. Although she has grown to like West Point, the commitment that comes at the beginning of junior year is a bit sobering.

Pete Haglin passed summer school chemistry and became even more committed to becoming an Army officer during Camp Buckner's military training. During yearling year he failed physics and returned to summer school. He was also caught off-limits—sneaking off-post at night—which earned him a serious punishment. Then he did the same thing again, got caught and punished again. He continues to count the days until he can exchange his cadet gray for Army green.

ACKNOWLEDGMENTS

West Point is a complex place, and my experience there, both as a cadet and as an instructor, is quickly fading into what the twenty-year-old subjects of this book would call "ancient history." I would like to thank the following people, who helped put modern West Point in perspective, and who allowed me generous access to their lives, their careers, and their insights into leadership.

Lieutenant General Daniel Christman, the fifty-fifth Superintendent, is justly proud of the cadets, staff, and faculty at West Point. That translated into his providing me unfettered access to every corner of cadet life and to the workings of the Academy. He set an example of openness, and the rest of the command followed his lead.

Thanks to Brigadier General John Abizaid, Colonel Maureen LeBoeuf, Colonel Joe Adamczyk, Majors Rob and Holly Olson, and Captain Brian Turner.

The USMA Public Affairs Office was generous in providing me information and a home-away-from-home. Thanks especially to

Andrea Hamburger, Major John Cornelio, Mike D'Aquino, and Theresa Brinkerhoff.

Thanks to all those who opened up their homes to me: John and Angela Calabro, Don and Jennifer Welch, and Kathy and Scott Snook.

Thanks to my generous friend and professional colleague, Paul McCarthy of McCarthy Creative Services. This book was Paul's idea; he entrusted it to me and coached me through its development. Thanks to Matt Bialer of the William Morris Agency, and to Henry Ferris of HarperCollins, who climbed on board midstream.

I owe the biggest debt to the cadets I had the pleasure of meeting and spending so much time with. They let me explore with them the meaning of their West Point experience, they let me watch them learn, and they shared with me their doubts and their triumphs. I especially want to thank Kevin Bradley and Kris Yagel, USMA 1999; Alisha Bryan, Grady Jett, Shannon Stein, and Greg Stitt, USMA 2000; Bob Friesema, Pete Haglin, and Jacquelyn Messel, USMA 2002.